Kepler's Geometrical Cosmology

Kepler's Geometrical Cosmology

J. V. FIELD

The University of Chicago Press

The University of Chicago Press, Chicago 60637
The Athlone Press, London

Printed in Great Britain

96 95 94 93 92 91 90 89 88 87 5 4 3 2 1

Library of Congress Cataloging-in-Publication Data

Field, J. V.
Kepler's geometrical cosmology.

Bibliography: p.
Includes index.
1. Cosmology. 2. Astronomy—History. 3. Kepler, Johannes, 1571–1630. I. Title.
QB981.F47 1987 512.1 87–10837
ISBN 0–226–24823–2

The publication of this work has been assisted by the support of the Royal Society.

To
Arthur Beer
(1900–1980)

Was kann der Mensch im Leben mehr gewinnen,
Als daß sich Gott-Natur ihm offenbare?
Wie sie das Feste läßt zu Geist verrinnen,
Wie sie das Geisterzeugte fest bewahre.

Contents

List of Figures

Acknowledgements for Figures

I am grateful to my colleague Miss P. F. Walker for drawing Figures 1.1–1.22, 2.1, 2.4, 2.5, 3.3, 4.1, 5.5, 5.15, A1.1, A1.2, A2.1–A2.3, A3.1–A3.3, A4.12–A4.14, A4.16 and A4.17.

Figures 2.2, 2.3, 3.14, 4.2, 5.6, 6.1 and A4.11 are reproduced by permission of the British Library; Figures 3.5, A4.6, A4.7, A4.10 and A4.15 by courtesy of the Biblioteca Ambrosiana, Milan; Figure 5.7 by courtesy of the Bayerische Staatsbibliothek, Munich; and Figures 3.1, 3.2, 3.4, 3.6–3.13, 5.1–5.4, 5.8–5.14, 5.16–5.18, 6.2, 6.3, A4.1–A4.5, A4.8 and A4.9 by courtesy of the Trustees of the Science Museum, London.

List of Tables

General Notes

1. All dates after 1582 are New Style unless otherwise stated.
2. All translations are by J. V. Field unless otherwise attributed.

Preface

As a deeply religious, if not quite orthodox, Lutheran and a convinced Platonist, Kepler saw the Universe as the outward expression of the nature of the Christian God. More radically, he interpreted Plato's *Timaeus* as a philosophical commentary on the book of Genesis. Furthermore, he saw the Copernican theory as providing him not only with the tools for rewriting the *Almagest* in accordance with Tycho Brahe's standards of accuracy and his own ideas of physics (*Astronomia Nova*, 1609) but also, once this task was accomplished, as enabling him to do the same for Ptolemy's *Harmonica*.

Like his astronomical results, Kepler's cosmological theories are products of his Copernicanism. They also show the working out of his Platonism in his professional practice as a *mathematicus*. This practice covered mathematics proper, astronomy, astrology and geometrical optics as well as cosmology and it is therefore to some of these fields of interest that we shall attempt to relate the following study of Kepler's cosmological theories. In particular, we shall not concern ourselves much with Kepler's physics (mainly Aristotelian in spirit) which does not seem to have made a direct contribution to the development of his cosmological theories, though it did, of course, contribute indirectly by influencing his astronomical work.

This book is based upon my Ph.D. thesis, accepted by the University of London in 1981. I am grateful to my supervisor, Professor A. R. Hall, for his perceptive and kindly guidance in the research that led to the thesis and his further aid and advice in the process of rewriting the work in its present form. I am grateful also to Dr E. J. Aiton for his unfailing generosity in helping and encouraging me in my research.

It was only after writing the Dedication that I learned that the lines I have quoted from Goethe's poem on Schiller's skull were quoted by Goethe himself as an ending to his novel *Wilhelm Meisters Wanderjahre*. I am glad to find that I am following the author's own

example in using them to express my gratitude for having learned to write.
June 1984

Additional Note

Relevant works which have been published since the completion of the typescript of this book are listed below in a supplementary bibliography. References to them have not been added to my text. I believe, however, that they tend to support the argument it presents.
March 1987

Introduction

The massive summing up of current research upon Kepler contained in the volume of *Vistas in Astronomy* (Beer and Beer 1975) that records the numerous international symposia which marked the four-hundredth anniversary of his birth, shows clearly that the greatest emphasis was still, as it always had been, upon Kepler's derivation of the three astronomical laws which bear his name. However, there were important studies of Kepler's philosophy and methodology by Buchdahl, Mittelstrass and Westman, presented to the symposium at Weil der Stadt in 1971, and a highly significant contribution to our understanding of Kepler's philosophy of science has since been made by Jardine (1984), in a book which developed from his translation of Kepler's unfinished tract in defence of Tycho against Ursus. Moreover, the most conspicuous imbalance between Kepler's own preoccupations and those of his historians has been partially redressed by the appearance of a full-length study of Kepler's theology (Hübner 1975). Unfortunately, we as yet still lack a modern edition of Kepler's theological works, as also of his astrological calendars.[1]

The present study is concerned with a part of Kepler's professional activity as a *mathematicus* which, in itself, has proved to be of less lasting significance than his astronomical work, namely his attempt to explain the structure of the Universe as a whole. Since Kepler was a devout Christian his cosmological theories are connected with his theological beliefs as well as with his astronomical theories, though it is the latter connection which is treated in detail in his cosmological works. Indeed, it is in a cosmological work, namely the *Mysterium Cosmographicum* (Tübingen, 1596), that Kepler gives his first analyses of the astronomical problems whose solution was to lead to the three laws, and some scholars, most notably Koyré (1961), have regarded the work in which the analyses are embedded as worthy of serious consideration. In fact, as Kepler himself noted in the second edition of the *Mysterium Cosmographicum* (Frankfurt, 1621), the form of these early analyses proved to be a fruitful one: 'almost every one of the astronomical works I have written since that time could be referred to some particular chapter of this little book, and be seen to contain either an illustration or a completion of what it says . . .'[2]

As it turned out, the astronomical analyses which led Kepler to his first two laws proved to be separable from the context of the architectonic theory in which they had first arisen, just as his third law proved to be separable from the rather similar work in which it first appears, namely the elaborate study of the mathematically harmonious nature of the Universe in *Harmonices Mundi Libri V* (Linz, 1619). There is thus a certain degree of justification for the positivist attitude of the numerous historians who have apparently been content to study the *Mysterium Cosmographicum* almost exclusively for its intimations of the *Astronomia Nova* that is to come, and to look upon *Harmonices Mundi Libri V* as no more than a rather antiquated setting for its one jewel, the third law. It is not, however, clear that there is any such justification for the concomitant assumption that the other elements in these two works are not worthy of serious attention – an assumption which modern historians may perhaps have inherited from some of Kepler's more influential seventeenth- and eighteenth-century readers (though we should also note that there were enough contemporary readers of the *Mysterium Cosmographicum* for the work to go into a second edition, at a date when its purely astronomical chapters had been superseded by the author's later works). We may note, also, that since Kepler's cosmological works, like his astronomical ones, require their readers to have a command of mathematics and astronomy most usually found only in a historian whose early training was in science, they have on the whole attracted scholars whose initial interest relates to that part of Kepler's work which they know from their earlier training, namely the three astronomical laws.

The neglect of Kepler's cosmological theories by the cosmologists of following generations must be at least partly attributed to the development of ideas about the Cosmos that is described by Koyré (1957) in *From the Closed World to the Infinite Universe*. In this development, Kepler stands out as a supporter of the older ideas, at least to the extent of believing that the Solar system was a uniquely important component of the Universe. Accordingly, both Koyré's and Kepler's arguments are reviewed in Chapter II below, since it is Kepler's belief in the special cosmological status of the Sun that made him see cosmological significance in a feature of the Copernican theory which Copernicus himself had treated rather casually, namely, the fact that since the Ptolemaic deferents of the inferior planets have now become reflections of the heliostatic orbit of the Earth, as have also the epicycles of the superior planets, it is possible to use astronomical methods to calculate the dimensions of the Copernican orbs (whereas the sizes of the Ptolemaic orbs were merely deduced from the physical assumption that they were arranged in a certain order and were all in contact with one another). Since Kepler's cosmological theories are largely concerned with explaining the dimensions of the planetary orbs, and later the velocities of the planets as seen from the Sun, it is clear that his Copernicanism is as fundamental to his cosmology as it is to his astronomy. Indeed, we know that Kepler himself regarded the polyhedral archetype described in the *Mysterium Cosmographicum* as confirmation of the Copernican hypotheses, for it was of this work that he wrote in 1598, 'It is enough honour for me that while Copernicus officiates at the high altar I can guard the door with my discovery'.[3]

It also appears that it was in pursuit of better astronomical evidence to support this cosmological confirmation of the Copernican model that Kepler accepted Tycho Brahe's invitation to work in Prague, for he wrote to Magini in 1601 'What influenced me most [in coming to Prague] was the hope of completing my study of the harmony of the world, something I have long meditated, and which I could only complete if Tycho were to rebuild Astronomy or if I could use his observations'.[4] Thus, if Kepler's cosmological theories exercised no perceptible influence on the development of cosmology, they did at least influence the development of Kepler's astronomy. Kepler himself used his new astronomical results not only for the traditional tasks of constructing tables and ephemerides, but also for the purpose he had mentioned to Magini, namely, to pursue his study of the mathematically harmonious structure of the Universe as a whole. Kepler's cosmology and his astronomy thus appear to be linked closely with one another, at least in Kepler's own mind. This fact in itself would seem to justify some further examination of the cosmological works, as a counterpart to the numerous investigations of the astronomical ones.

We have already noted that Kepler's cosmological works seem to have exercised little influence, though the *Mysterium Cosmographicum* appears to have been well received at the time of its first publication, in 1596, and went into a second edition in 1621 (thus providing the historian with valuable evidence of the author's changes of opinion over the intervening years – see Chapters III and IV below). Nevertheless, we know that Kepler himself thought well of both works, as may be seen from his enthusiastic introduction to *Harmonices Mundi* Book V (see Chapter VI below), and from the fact that he allowed a second edition of the *Mysterium Cosmographicum* to be printed. Kepler is not famous for the sureness of his historical judgement, but it appears that in this case historians have been too ready to neglect his opinion. He does indeed seem to have believed his cosmological works would stand the test of time, and on this point we must, with hindsight, disagree with him. There remains, however, his judgement that the *Mysterium Cosmographicum* and *Harmonices Mundi Libri V* were good examples of natural philosophy. It therefore seems worthwhile to ask ourselves what they tell us about Kepler's attitude to the study of nature, and to compare this with the evidence we find in some of the other works for which he is now generally better remembered. The present study will thus be concerned with setting Kepler's cosmological works in the context of his work as a whole, but most particularly with examining their relation to his astronomical works, since it was crucial to the nature of Kepler's cosmological theories that they were designed to explain the relations between magnitudes which could be calculated from the results of astronomical observation.

Relations between magnitudes were the province of geometry, so it was natural that Kepler should cast his cosmological theories in geometrical form. However, he did much more than that: he turned to geometry itself (that is, to the nature of Euclidean space) as providing an explanation for the structure of the Universe. Plato is thus as important an influence as Copernicus, and we shall therefore briefly consider some aspects of *Timaeus* before attempting an analysis of Kepler's contributions to the tradition it represents.

I

Platonic Science

Both Kepler's principal cosmological works, the *Mysterium Cosmographicum* (Tübingen, 1596) and *Harmonices Mundi Libri V* (Linz, 1619), are deeply indebted to Plato's *Timaeus* and to Proclus' *Commentary on the First Book of Euclid's Elements.* A marginal note on a line in a long quotation from the latter work which happens to refer to the former one may serve to exemplify Kepler's attitude: '[This is to be found] in *Timaeus*, which is, beyond all possible doubt, a commentary on the book of Genesis, otherwise the first book of Moses, transforming it into Pythagorean philosophy, as will easily be apparent to an attentive reader who compares it with Moses' own words'.[1] (There can be little doubt that Kepler wrote his own marginalia: the next but one reads 'Puto 1.2.4.8.3.9.27', which is hardly the kind of remark one might expect to originate with the printer.) Since Kepler's judgement of *Timaeus* may seem to be extreme to the point of rashness it should perhaps be pointed out that Kepler, like many another reader, believed that *Timaeus* was not necessarily to be taken entirely at its face value. For example, in *Harmonices Mundi* Book III Chapter I he quotes Timaeus' suggestion that harmonic proportions were to be found in the structure of the soul, adding that Aristotle had refuted the argument presented in *Timaeus*, but remarking that he, Kepler, believed Plato's words should not necessarily be taken literally (KGW *6*, p. 107). He must have reasoned similarly to account for the fact that *Timaeus*, which he correctly recognised as indebted to Pythagorean theories, appeared to describe a geocentric Universe, whereas he not only believed the Universe to be heliocentric but also ascribed this opinion to the Pythagoreans.

The scope of Kepler's *Harmonices Mundi Libri V* is both larger and smaller than the scope of Plato's *Timaeus.* It is larger in that Kepler sets out to explain the details of the structure of the system of planets whereas Plato had only given a diagrammatic, and at the same time very poetic, description of the heavens, which, as Cornford has remarked, rather conveys the impression of an armillary sphere informed by soul. Plato's astronomy is perfectly satisfying, as far as it goes, and his breaking down of celestial motions into two components, rotation about the celestial poles and motion along the ecliptic, was a beautiful demonstration of the power of mathematics to expose the simplicity that lies behind the complicated world of appearances, that is, in Plato's terms, its power to turn our eyes from the flickering shadows on the wall of the cave and direct them instead to the models that cast the shadows.

The scope of the *Harmonice Mundi* is narrower than that of *Timaeus* in that while Plato sets out to give a fairly detailed, though not very full, account of sublunary phenomena, for example by discussing various properties of the four sublunary elements, Kepler contents himself with a more sketchy account, expressed in terms of the response of souls to the harmonic ratios in which heavenly circles are divided when celestial bodies are at certain Aspects to one another. Plato, in fact, gives us the outlines of a detailed mechanistic theory of the properties of matter, whereas Kepler provides no more than a mechanistic cum animistic astrology, which does not yield a detailed account of sublunary phenomena. It is, in fact, apparent from Kepler's other writings on astrology, for example *De Fundamentis Astrologiae Certioribus* (Prague, 1602), that he did not believe astrology could ever yield a detailed account of such phenomena, so we must assume that Kepler was aware that the explanation of the Universe given in *Harmonices Mundi Libri V* was, to this extent, incomplete.[2] Kepler's main concern was, in any case, always with astronomy and it appears that he never concerned himself directly with the properties of the elements. His nearest approach to it was the very short and highly mathematical *De Nive Sexangula* (Prague, 1611), which entirely fails to explain the shape of snow, though it gives an admirable account of the shape of the honeycomb.

The theory of matter described in Plato's *Timaeus* is the only more or less fully worked out scientific theory described in Plato's work: the one specimen of autograph Platonic science. Although the theory is presented by Timaeus of Locri rather than by Socrates it is clear that Plato at least considered it worthy of serious attention and for our present purposes we may characterise it merely as 'Platonic' rather than seek to disentangle the components which should be ascribed to the Pythagoreans, Empedocles, the Atomists, Timaeus of Locri personally and, perhaps, even Plato himself. This attitude is, in principle, unhistorical, and what follows will not purport to be a serious historical account of the opinions held by Plato. Our concern is narrower: to see what kind of scientific theory Plato describes. We are thus, as it were, presenting a possible 1590s reading of *Timaeus*.

Plato does not describe his God as creating the world out of nothingness, but rather as making order, a cosmos, out of disorder. That is, the elements are all already present when Plato begins his account, and he is thus concerned with their properties rather than their origin.

The theory of matter of *Timaeus* is described entirely in terms of geometrical properties and geometrical relationships, that is, in terms which can only be applied to mathematical entities, which Plato regarded as belonging to the realm of the 'forms'.[3] That the polyhedral figures of the elements are indeed to be considered as mathematical entities is confirmed by the fact that Timaeus describes them as resulting from forms and numbers (53b). However, it is never quite clear what relationship the figures bear to the real particles we might find if we made a sufficiently delicate analysis of, say, a lump of clay. We are told (56c) that all the elements or seeds are to be thought of as being too small for us to see them individually, and it is quite possible that Plato would have considered it meaningless to ask whether the seeds of earth would appear cubic if we could see them (the exact relationship between observable bodies and Platonic 'forms' is a matter of considerable dispute). Moreover, it should be noted that, despite the a priori arguments which Timaeus presents, he states explicitly that his theory is merely probable, adding that in any case only a god or someone specially favoured by that god could know the higher causes which explain why particular figures should be involved (53d).

The basic figures: triangles

Timaeus begins by pointing out that since the elements are bodies their forms must include thickness, which involves having surfaces (53c). It then appears that these surfaces are to be 'straight' (ὀρθος) and we are told that a 'straight' surface is always made up of triangles. The English word 'straight' is clearly an inadequate translation of Plato's ὀρθος for the purpose of this passage, but the line of thought is not difficult to reconstruct: Timaeus is assuming that the surfaces concerned will be plane, and since two planes intersect in a line, the edges of the solid figure will be straight, i.e. the faces of the solid figure will be polygonal. Any polygon can, indeed, as he says, be dissected into triangles (see figure 1.1)

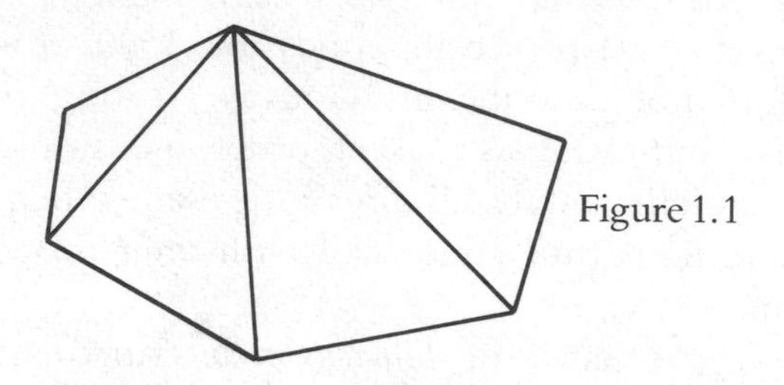

Figure 1.1

Timaeus next states that 'all the triangles' arise from combinations of two kinds of triangle, both of them right-angled. One of the basic right-angled triangles is isosceles, the other is not (it will eventually prove to be the triangle whose angles are 90°, 60° and 30°). It is true that any triangle can be dissected into two right-angled triangles: we only need to drop a perpendicular from one of the vertices on to the opposite side (see figure 1.2).

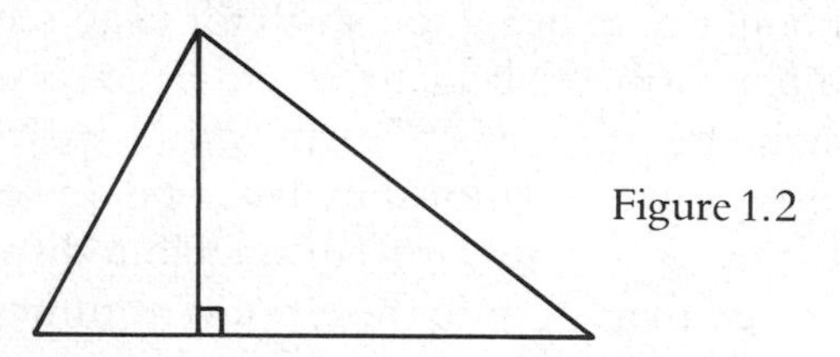

Figure 1.2

However, in the general case the right-angled triangles produced in this way will not be of either of the special kinds that Timaeus goes on to consider. It therefore appears that 'all the triangles' (τὰ δὲ τρίγωνα πάντα) must refer only to all the special triangles Timaeus is going to consider in the theory that follows, that is, 90°, 45°, 45° triangles and 90°, 60°, 30° triangles. If we take 'all the triangles' in this sense, what Timaeus says is true: a 90°, 45° 45° triangle can be dissected indefinitely into more 90°,

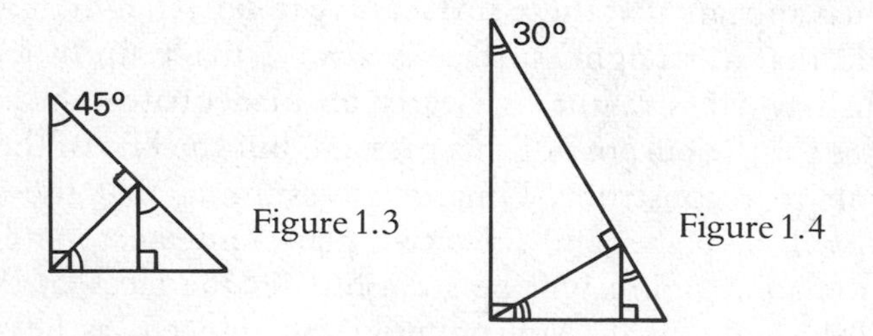

Figure 1.3

Figure 1.4

45°, 45° triangles and a 90°, 60°, 30° triangle can be dissected indefinitely into more 90°, 60°, 30° triangles (see figures 1.3 and 1.4).

We are not offered any particular justification for the assumption that one of the basic triangles must be isosceles, but Timaeus notes that while the isosceles right-angled triangle is of only one kind (μίαν εἴληχεν φύσιν) the non-isosceles one can be of any one of an unlimited number of kinds (54a), and we must therefore simply choose the most beautiful. In earlier passages of the dialogue it has already been agreed that God's actions were undoubtedly intended to make the Cosmos as beautiful as possible so the reappearance of the aesthetic principle at this point cannot be regarded as unreasonable. However, Timaeus declines to reason as to why the particular non-isosceles triangle he chooses should be considered the most beautiful. In fact, he says that if anyone can find a more beautiful triangle than this one then he, Timaeus, will welcome him as a friend, but that meanwhile we shall choose the triangle which is half an equilateral triangle. The following sentence is somewhat obscure. What it says seems to be that it would take too long to explain this decision but if anyone can give a demonstration that it is justified (or '*not* justified' – some manuscripts apparently have an additional μὴ) he will be welcome to the prize. It is in any case clear that Timaeus is not pretending that he has justified his choice of this one particular kind of non-isosceles right-angled triangle on a priori grounds. He goes on to say that although the four elements appear to be generated from one another they are in fact generated from these triangles (54c), three from one kind (the non-isosceles triangle) and only one from the other (the isosceles triangle). Therefore it is not possible for the fourth element to be generated from the other three, nor vice versa.

He then turns to the forms of the individual elements, starting with those put together from the non-isosceles triangles. He describes how six such triangles can be put together to form an equilateral triangle (54d–e, see figure 1.5).

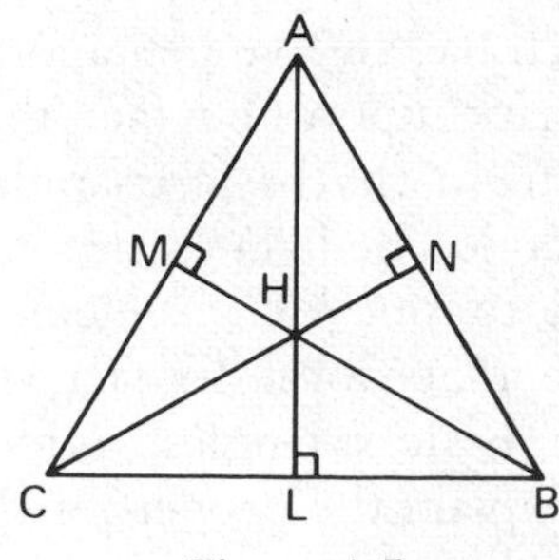

Figure 1.5

Four such equilateral triangles make a tetrahedron, eight an octahedron and twenty an icosahedron (figures 1.6, 1.7 and 1.8, 54e–55b). The isosceles right-angled triangle will construct the

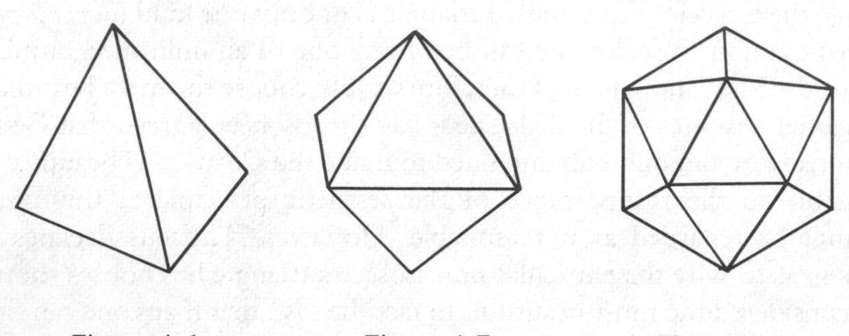

Figure 1.6 Tetrahedron

Figure 1.7 Octahedron

Figure 1.8 Icosahedron

cube, four triangles being fitted together to form each face (figure 1.9) and six faces to form the solid (figure 1.10, 55b–c).

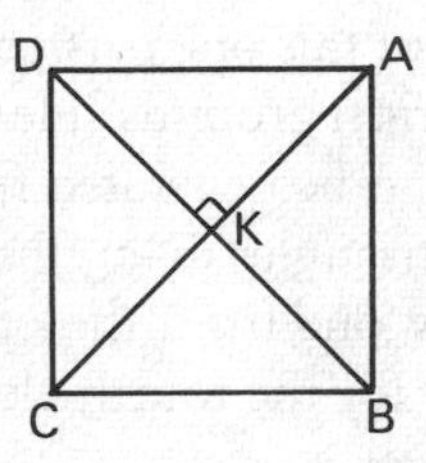

Figure 1.9

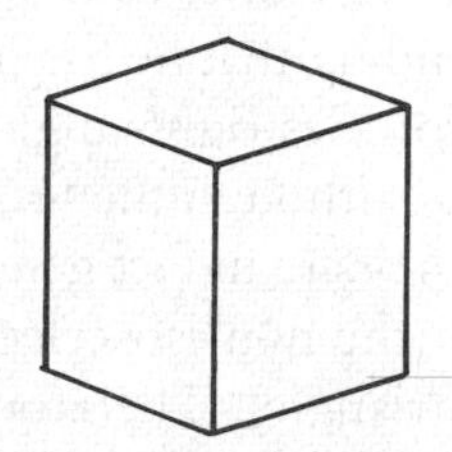

Figure 1.10 Cube

Timaeus then adds that there remains just one more construction (συστασις), the fifth, which God used in designing the whole. This final remark must refer to the last of the convex regular polyhedra, the dodecahedron, which has twelve pentagonal faces (figure 1.11). Timaeus does not remark upon the fact that the dodecahedron cannot be constructed from the basic triangles used for the other four figures, and the dodecahedron is not mentioned when each polyhedron is assigned to one of the elements in the following section of the dialogue (55e–56b). In fact, Plato's main concern seems to be with the basic triangles, which he uses to make polyhedra and thus to explain the properties of the elements, rather than with the regular polyhedra themselves.

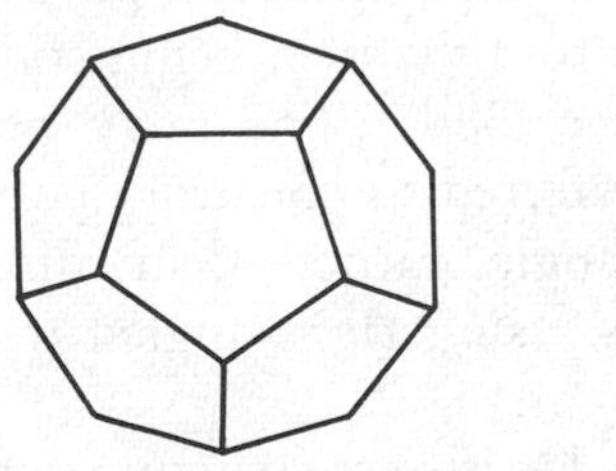

Figure 1.11
Dodecahedron

As we have seen, Timaeus has justified his use of triangles on the grounds that triangles are the most basic of figures (53c–d) and has made it clear that he believes there are reasons for preferring the particular triangles he has chosen (54a–b). However, he nowhere explains why he chooses to assemble the equilateral triangles he requires in the rather elaborate manner shown in figure 1.5 instead of in the simpler pattern shown in figure 1.12. Nor does he explain why

Figure 1.12 Figure 1.13

the squares are assembled as in figure 1.9 rather than as in figure 1.13. It seems likely that one explanation for the use of the patterns shown in figures 1.5 and 1.9 is the fact that these patterns involve grouping triangles round the centre of the figure (H in figure 1.5, K in figure 1.9). Plato is certainly aware that the tetrahedron can be inscribed in a sphere, since he mentions the fact at 55a, and it is to be presumed that he realised that the same applied to the other regular polyhedra. He must also have known that the faces of the solids had an analogous property, namely that they could be inscribed in a circle. What has been called the 'centre' of the equilateral triangle and the square is the centre of the circle in which each of the figures could be inscribed. The reasoning ascribed to Plato might be expressed less rigorously by saying that the patterns in figures 1.5 and 1.9 are 'more symmetrical' than those in figures 1.12 and 1.13.

When we come to the application of the theory, we find that we are required to construct elementary polyhedra in different sizes. For example, we are told that ice, being solid, is made up of larger polyhedra than water (58d). This may perhaps suggest another reason why Plato preferred to construct polyhedra from two kinds of right-angled triangle rather than directly from equilateral triangles and squares: using right-angled triangles allows a greater variety of size.

If we are concerned with constructing square faces, the use of a square unit will give us the series shown in figure 1.14. The areas of

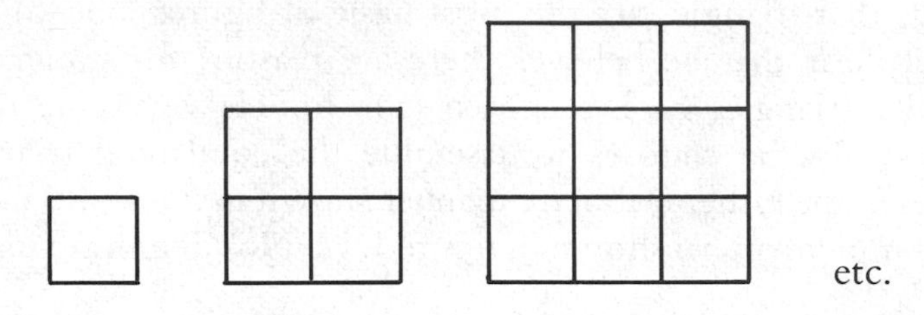

Figure 1.14

the faces are clearly going to be 1, 4, 9, 16, 25, etc., the area of the nth face being n^2. The use of an isosceles right-angled triangle as a half-unit building block will give us two series of faces. The first begins with the square shown in figure 1.15, which gives us a series identical with that shown in figure 1.14, but we can also construct the square shown in figure 1.16, which will give us a series whose nth term is $2n^2$. The use of the triangle rather than the square has

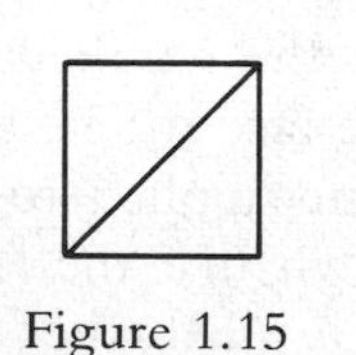

Figure 1.15

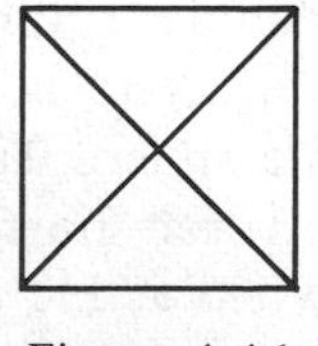

Figure 1.16

thus enabled us to construct twice as many sizes of face for our cubes.

If we are concerned with constructing faces which are equilateral triangles, the use of an equilateral triangle unit will give us the series shown in figure 1.17. The areas of the faces are clearly going

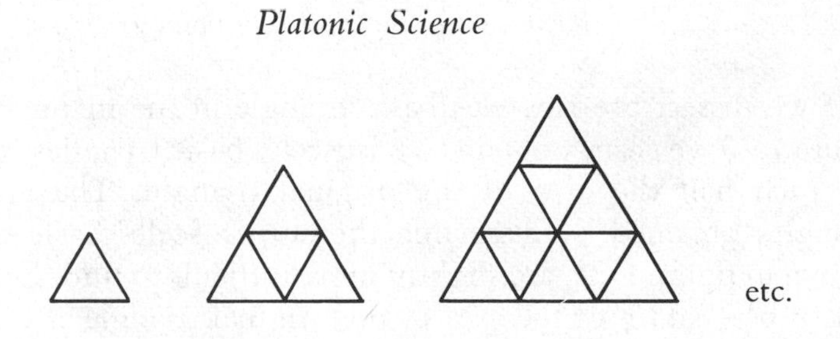

Figure 1.17

to be 1, 4, 9, 16, 25, etc., the area of the *n*th face being n^2. The use of the 90°, 60°, 30° triangle as a half-unit building block will give us two series of faces. The first begins with the triangle shown in figure 1.18, which gives us a series identical with that shown in figure 1.17, but we can also construct the equilateral triangle shown in figure 1.19, which will give us a series whose *n*th term is $3n^2$. The use of the 90°, 60°, 30° triangle rather than the equilateral triangle has thus enabled us to construct twice as many sizes of face for our tetrahedra, octahedra and icosahedra.

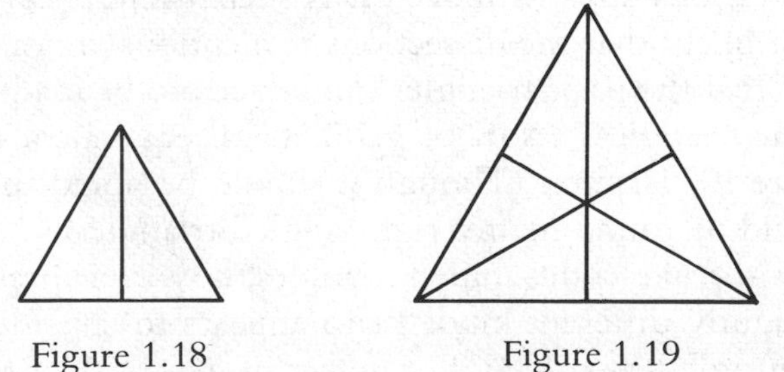

Figure 1.18 Figure 1.19

It is, however, also possible that Plato envisaged the different sizes of polyhedra as being put together from basic triangles of different sizes. This might explain the remark at the end of 53c that all the basic triangles can be seen as made up of smaller basic triangles (see figures 1.3 and 1.4, and figures 1.20 and 1.21).

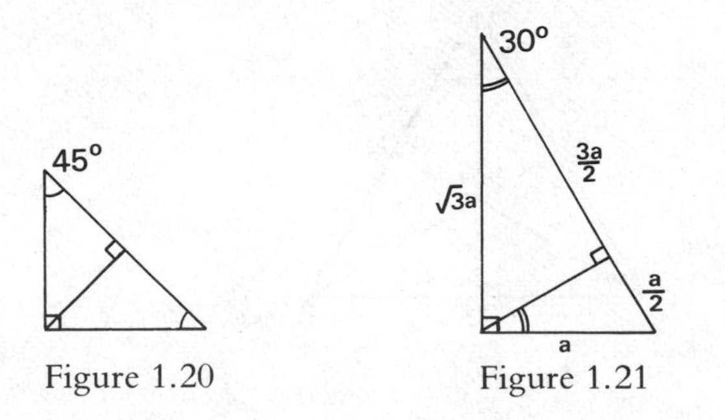

Figure 1.20 Figure 1.21

If we dissect the isosceles basic triangle in the manner shown in figure 1.20 we clearly obtain two isosceles basic triangles whose areas are each half the area of the original triangle. The areas of the triangles obtained by dissecting the non-isosceles basic triangle, as shown in figure 1.21, are slightly more difficult to find, but they turn out to be $\frac{1}{4}$ and $\frac{3}{4}$ of the area of the original triangle.

If we now apply the constructions shown in figures 1.14, 1.15 and 1.16 using a basic triangle of area $\frac{1}{4}$ rather than $\frac{1}{2}$, we shall obtain series of areas whose general terms are $\frac{1}{2}n^2$ and n^2. Some of these areas are different from those we obtained before. If we apply the constructions shown in figures 1.17, 1.18 and 1.19 using the smaller basic triangles shown in figure 1.21 we shall again find that some of the areas we obtain are different from those we obtained before.

It is clear that repeated dissections of the basic triangles, as shown in figures 1.20 and 1.21, will continue to yield more basic triangles and more possible areas for faces. However, the remark at 53c appears to be the only reference to the possibility that basic triangles may be of various sizes (if indeed it is such a reference) and it thus seems most likely that the dissections it mentions are only intended to establish that the basic triangles can be seen to be made up only of similar basic triangles, a sort of geometrical equivalent to showing that they are 100% pure. (Though it should be noted that the same purity would be found in any right-angled triangle.)

The non-isosceles right-angled triangles, however, present a problem of a slightly different kind. Plato appears to consider that only one such triangle is involved, but in fact, if we consider the triangles as purely mathematical entities, confined to their own plane, then we find that the construction of the equilateral triangle described in 54d–e requires two different kinds of non-isosceles right-angled triangle (see figure 1.22). In this figure we can see that triangle HBL

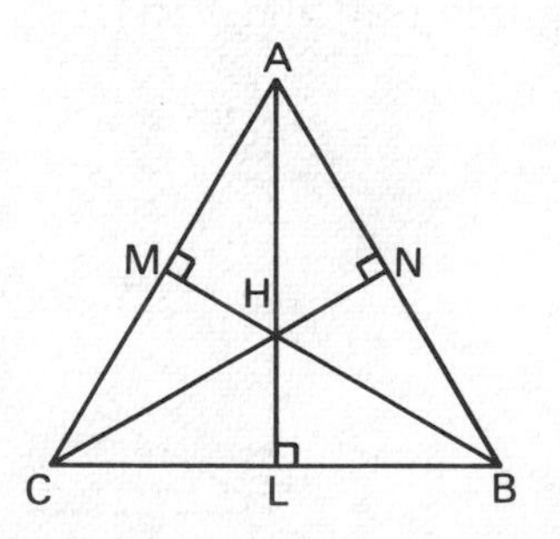

Figure 1.22

is indeed congruent with triangle HLC in the modern sense, that is, all corresponding sides and angles are of equal size, but the two triangles are mirror images of one another. The usual definition of geometrical equality is that expressed in Euclid's Common Notion 4,[4] namely that things which coincide with one another (or can be superimposed upon one another) are equal. As Heath notes, the way Euclid uses this notion, for example in Book I proposition 4 and Book I proposition 8, 'leaves no room for doubt that he regarded one figure as actually *moved* and *placed upon* the other'. Heath concludes that Euclid's apparent distaste for the method of superposition probably indicates that the method was a traditional one. He adds that it may well date back to Thales. Now it is clear that, in figure 1.22, triangle HBL cannot be superimposed on triangle HLC unless we can, as it were, allow ourselves to pick it up and turn it over, that is, unless we regard the triangles not as being confined to the plane, as purely mathematical entities, but as having a physical existence as two-dimensional bodies in three-dimensional space.

Euclid never mentions the possibility of figures being mirror images of one another, though his avoidance of the method of superposition on a number of occasions when it might have seemed natural to think of employing it, for example in Book I proposition 26, may perhaps indicate that the idea of mirror images had occurred to him. Proclus' comment on this proposition unfortunately deals mainly with the form of the enunciation and not at all with Euclid's choice of the method of proof. In any case, we do know that the idea of mirror images was used by Pappus, in an alternative proof for the theorem which is proved in Proposition 5 of Book I of the *Elements*. Our account of this proof is due to Proclus, who presents it as originating with Pappus. Heath quotes this proof in his note on Book I Proposition 5 but makes no comment on its connection with the idea of mirror images.[5]

Since Plato appears to believe that only one kind of basic triangle is involved in the construction shown in figure 1.22, I think we must assume that he envisaged the triangles as behaving as if they were two-dimensional objects free to move in three-dimensional space. In this respect Plato's God seems to be a faintly Aristotelian geometer.

The figures of the elements: polyhedra

Having described how the tetrahedron, the octahedron, the icosahedron and the cube can be constructed from the basic triangles (in 54d–55d), Timaeus proceeds to assign each polyhedron to an element (55e).

He makes the most straightforward attribution first: 'We shall surely give earth the figure of the cube' (55e). Timaeus' justification for the decision is as follows: Earth is the most difficult of the elements to set in motion, and must thus have the most stable type of face. (The same word, βασις, is used to mean either 'face' or 'base' and any of the polyhedra can, of course, stand on any of its faces.) The more stable face is that made up of isosceles triangles, therefore the square face is both in its parts and as a whole more stable than the triangular one (55e).

There cannot be any doubt that this argument is concerned only with the shape of the face and not with the three-dimensional shape of the cube. Moreover, the stability of the face appears to be ascribed directly to the stability (that is, presumably, the symmetry) of the isosceles triangles which are its ultimate indivisible components.

When we turn to the other three elements the attributions must naturally take account of the solid shapes of the polyhedra concerned, since all their faces are exactly the same, being equilateral triangles. Timaeus makes these attributions by considering firstly how easily each of the figures can be moved, secondly how large each is and thirdly the sharpness of its solid angles.

The meaning of the first consideration is not clear from *Timaeus*, but a passage in *On the Heavens* Book III Chapter I (229 b 33) suggests that Aristotle had interpreted Plato as referring merely to the weight of the polyhedra, which, since they are made up of surfaces, will depend upon the surface area. If the faces are the same size in every figure, the lightest figure will be the tetrahedron and the heaviest the icosahedron, as Plato apparently requires. It is far from obvious that Aristotle can in general be relied upon as a faithful guide to the meaning of *Timaeus* but in this case his interpretation does seem to be acceptable. (The 'obvious' interpretation of 'easy to move' as meaning that the polyhedron would roll easily does not seem to be tenable, since the icosahedron, the most nearly spherical of the figures, is pronounced the 'most difficult to move'.) If we assume that 'easy to move' does, indeed, mean 'light', then the fact that we know that the particles of earth are the heaviest ones of all tells us that the weight of 24 (that is 6 × 4) isosceles basic triangles must be greater than the weight of 120 (that is 20 × 6) non-isosceles basic triangles, that is, one isosceles basic triangle is heavier than five non-isosceles basic triangles. This is all we are ever able to establish as to the relative sizes of the two types of basic triangle.

If we again assume that we are to compare tetrahedra, octahedra and icosahedra with faces of the same size, the second consideration, the size of the figures, may be supposed to refer to their overall diameters or to their volumes. Since Timaeus appears to regard it as merely obvious that the tetrahedron is the smallest figure and the icosahedron the largest, we have no way of deciding exactly what he means.

The third consideration, the sharpness of the solid angles, presents no problems, though it should be noted that it is only later that we are explicitly given grounds for believing, for example, that the points of the seeds of fire are sharper than those of the seeds of water. The observational evidence for the attributions is adduced only when we come to applications of the theory.

All three considerations happily suggest exactly the same attributions: the tetrahedron for fire, the octahedron for air and the icosahedron for water (56b).

Transmutation of the elements

As we have seen, the cube is composed of isosceles right-angled triangles, which are peculiar to the cube, so we cannot expect the element earth to be transformed into any other element. However, fire or air or water can break up earth, and the broken pieces may eventually reassemble themselves into earth, though never into another element (55d).

The other three elements can be transformed into one another, and Timaeus provides several examples for such transformations (56d–e). The following are typical.

1. If water is broken up by air or fire, the parts can come together again to form one particle of fire and two of air. That is, in terms of the triangular faces, $20 \rightarrow (1\times4)+(2\times8)$; and in terms of basic triangles, $20\times6 \rightarrow (1\times(4\times6))+(2\times(8\times6))$.
2. Air can be divided up to give two particles of fire. In terms of faces we have $8 \rightarrow 2\times4$; in terms of basic triangles, $8\times6 \rightarrow 2\times(4\times6)$.
3. Fire, if trapped by masses of water or earth, can be crushed in such a way that two particles of fire give one of air. In terms of faces we have $2\times4 \rightarrow 1\times8$; in terms of basic triangles, $2\times(4\times6) \rightarrow 1\times(8\times6)$.
4. Air, similarly crushed, can be converted into water: two and a half (*sic*) particles of air giving one of water. In terms of faces we have $2\frac{1}{2}\times8 \rightarrow 1\times20$; in terms of basic triangles, $2\frac{1}{2}\times(8\times6) \rightarrow 1\times(20\times6)$.

All these transformations are described as resulting from physical contact between polyhedral particles. For example, the sharp points of fire tear apart the particles of water, or the heavy masses of water crush the particles of fire. Nothing in this section makes it clear whether we are to suppose the polyhedra to be broken down into faces or into basic triangles, though the latter seems more likely, both because the triangles are basic and because it is easier, for example, to imagine the sharp point of a tetrahedron breaking up a face of an icosahedron rather than prising the icosahedron apart along one of its edges. However, it is perfectly clear that dissolutions and recombinations are to be seen in terms of triangles of some kind and not in terms of cutting up solid polyhedra, as is supposed by Rivaud in the introduction to his translation of Timaeus.[6]

Applications of the theory

The theory outlined above allows Timaeus to give mechanistic explanations for various phenomena. For example, the fact that air may be either the very pure kind called aether or the murky kind known as mist is explained as being due to the various sizes of the triangles concerned (58d). The freezing of water is explained similarly, and fire particles are said to be slowly forced out as the water cools and becomes more condensed (59a). Metals are said to be a variety of water which is solid except when it is made very hot (59b), that is, in modern terms, metals are thought of as varieties of water which have very high freezing points. There are also explanations of the different varieties of ice: as hail, snow and hoar frost (59e).

Timaeus eventually turns to explaining how the particles affect the human senses (61c) and gives, among other things, a relatively detailed explanation of the sensation of smell (66d). This explanation seems to typify the way in which Plato expects his theory to account for natural phenomena, so we shall attempt to summarise it.

There are, Timaeus says, no particular kinds of particle that affect the organs of smell, and there are no particular particles that correspond to particular smells, sweet smells, sour smells and so on. However, the vessels that take in smells are too narrow for particles of earth or water to pass through, but too wide to be affected by particles of fire or air. So none of these elements has a smell that anyone has ever noticed. Smells only arise when certain bodies get wet, rot, turn into liquid or evaporate. Indeed it is when water is turning into air or air into water that smells arise, being caused by the bodies of intermediate size that are produced in the course of the transformation of one element into the other (66d–e).

Now, we have already been told, several times, that particles of any element come in many different sizes – this was, for example, how Timaeus accounted for the various types of air (58d). Therefore the suggestion that the passages which admit smells are too small for the particles of some elements and too large to be affected by the particles of others is not, on its own, satisfactory. Moreover, although we may perhaps compare the sizes of particles of fire, air and water, which are generated from one another and make use of the same basic triangles, we have not been given any way of comparing these particles with those of earth, which use a different basic triangle, whose size (in the sense of area, side, perimeter or anything else) may well be different from that of the triangles used for the other elements. Indeed, as we saw in connection with the attribution of the polyhedra to the elements (55e–56b), it does appear that Plato intended that the triangles used for the cube should be heavier than those used for the other polyhedra, assuming the square and triangular faces were to be made up in the way he describes, using four triangles for the square and six for the equilateral triangle. It is not only that a great deal is being left as an exercise for the reader's ingenuity, but also that the theory is essentially incomplete, in the form in which Timaeus presents it. However, if we assume that air with faces of some given size is in the process of turning into water with faces of the same size, then we may, indeed, imagine that intermediate bodies formed will be intermediate in size between the particles of air and of water.

All of Timaeus' explanations seem to be of this kind: outlines that might be regarded as providing an indication of the kind of explanation that could perhaps be constructed. Moreover, the outlines are all mathematical, and therefore relate to the 'forms', not to the details of observed appearances, the insubstantial details of the shadows on the wall of the cave, shadows of which we can have only 'opinion', not 'knowledge'.[7] In fact, the scientific theory presented in the *Timaeus* seems to fit in quite well with the theory of knowledge developed in the *Republic*, which was probably written at least ten years earlier, but which Plato apparently wishes us to imagine as having taken place two days before the dialogue recorded in *Timaeus*, the latter dialogue being concerned to place the ideal society described in the former in its cosmographical context, and to assert, in the story of Atlantis, that such an ideal society did once really exist in Athens.

Kepler's use of *Timaeus*

Kepler shared with Plato the belief that the Universe was in some sense an expression of the nature of the God (or, more probably, in Plato's case, gods) who created it. They also shared the belief that the creator was a geometer. However, what Kepler demanded of a scientific theory was very different from what Plato demanded and there does not seem to be much room for doubt that Kepler rejected the theory of matter described in the previous sections of this chapter. Certainly, in *De Nive Sexangula* (Prague, 1611) he never even mentions polyhedral elementary particles, not even as a possible way of accounting for, say, the action of cold air on water to form snow. However, Kepler was very far from rejecting the method Plato had employed in arriving at his theory. As we shall see in later chapters, Kepler's use of mathematics in his cosmological works is very like Plato's use of mathematics in *Timaeus*. However, unlike Plato, Kepler constructs not a vague metaphysical description of the cosmos but a testable mathematical model: Plato's polyhedral theory of the elements is designed to deal with unquantifiable properties, whereas Kepler's polyhedral theory of the Solar system is concerned with the measurable intervals between the planetary orbs.

If we turn from Kepler's actual practice to his account of its philosophical basis (that is, if we turn from what Kepler does to what he says he is doing), we find that he appeals to the philosophy of mathematics developed by Proclus, that is, again, indirectly, but not very indirectly, to Plato. The details of Kepler's geometry are, of course, mainly derived from Euclid.

II
The Size of the Universe and the Status of the Sun

In a geocentric cosmology the Sun's motion is closely related to the motion of the other planets, but in a heliocentric cosmology the Sun shares the immobility of the fixed stars. Whether one is then prepared to take the further step of asserting that the Sun has no special status but is simply another star proves to depend at least partly on whether one is willing to accept the consequence that the distance from the Sun to the nearest star must be very large indeed compared with the distance between the Sun and the Earth (far larger than would be required merely to ensure that annual parallaxes would be less than, say, 1′ of arc). Some of Kepler's contemporaries, among them Giordano Bruno and William Gilbert, not only accepted this consequence but went on to assert that the Universe was infinite.

That Kepler did not share this view is apparent from the two works with which we are mainly concerned: the *Mysterium Cosmographicum* (Tübingen, 1596) and the *Harmonice Mundi* (Linz, 1619). The titles proclaim both works to be concerned with the structure of the Universe yet each deals only with the Solar system. Neither work even discusses the status of the Sun, though we know that by 1606 Kepler knew at least something of the cosmological work of Bruno and of Gilbert, since in *De Stella Nova* (Prague, 1606) he mentions their belief that the Universe is infinite (KGW *1*, p. 253, l.4). The reason for his silence in the *Mysterium Cosmographicum* and in the *Harmonice Mundi* is not far to seek: Kepler was in agreement with the majority of his contemporaries in believing that the Universe was finite and that the Sun held a privileged position. However, since this belief is crucial to the theories put forward in the *Mysterium Cosmographicum* and in the *Harmonice Mundi* it is of interest to examine the arguments Kepler adduces in support of it in other works. Koyré (1957) mentions that Kepler had metaphysical reasons for denying the infinity of the Universe and adds that these reasons were derived from his religious beliefs.[1] It is indeed true that Kepler sees the Sun, the stars and the intervening aether as manifesting the Trinity of the Father, the Son and the Holy Ghost[2] and this analogy would clearly be entirely unsatisfactory if the Sun were no more than a typical star, as in Bruno's system. It should, however, be noted that when Kepler discusses the problem of the size of the Universe, in *De Stella Nova* (Prague, 1606) and in the *Epitome Astronomiae Copernicanae*, Book I (Linz, 1618), he does not mention his analogy at all, not even when, in the earlier work, he admits that he recoils from the idea of an infinite Universe: 'The mere thought brings with it I know not what of secret horror, as one finds oneself wandering in this immensity, whose boundaries, whose centre, and thus any defined place at all, are all denied to exist.'[3] The analogy may, of course, have weighed with him in private, but he never, as far as I know, presented it as an argument for others.

Kepler's philosophical argument against the idea of an infinite Universe is directed against the limitations it would place upon the value of observation. If the Universe is immense or infinite, in the sense that we must believe it to contain objects we cannot observe, such as stars that are sometimes too far away for us to see them (such stars had been invoked to explain the observation of 'new' stars in 1572 and 1604), then we cannot know how to construct theories to explain what we observe. In the particular cases of the theories just mentioned Kepler is merely objecting to the cosmological equivalent of admitting as acceptable a theory that a forest fire might have been started by a dragon which wandered in from somewhere over the edge of the map. His more general demand, in modern terms, is that the word 'Universe' must be taken to mean 'observable Universe'. It is hardly surprising that he finds it possible to show that Aristotle is on his side.[4]

However, Aristotle's proof that the Universe is finite will not serve Kepler's turn, since it depended on the motion of the sphere of fixed stars and, as Kepler says in *De Stella Nova*: 'Copernicus, having removed the motion of the sphere of the fixed stars, would allow it to be infinite'.[5] Since Kepler is justly famed for misrepresenting Copernicus' ideas it may be as well to point out that on this occasion he is not doing so. In *De Revolutionibus* Copernicus writes: 'For their strongest argument that the world is finite is motion. So let us leave it to the natural philosophers to debate whether the world be finite or infinite . . .'[6]

We shall not concern ourselves further with the entirety of

Kepler's proof that the Universe is finite. This has been dealt with in some detail by Koyré in *From the Closed World to the Infinite Universe* and for our present purposes there is very little to add to his discussion of the larger issues. Our concern here is rather with Kepler's proof that special status must be accorded to the Sun. Kepler deals with this subject in two very different works: in *De Stella Nova* (1606), a learned treatise on the new star of 1604, and in *Epitome Astronomiae Copernicanae* (1618–21), a textbook which is designed to cover the whole subject of astronomy and is therefore a valuable guide to Kepler's opinions at this time (particularly since he provided it with an index).

De Stella Nova (1606)

The new star which was first seen on 9 October 1604 is now generally known as 'Kepler's supernova', though Kepler was not the first astronomer to observe the star but merely learned of its appearance by report – the weather at Prague prevented him from seeing it until 17 October. (No doubt Kepler regarded this as evidence to confirm his belief that one of the astrological consequences to be expected from the appearance of the star was an increase in rainfall.) At this time, Kepler was deep in his work on the orbit of Mars, but he found time to write a pamphlet on the new star, *Bericht vom Newen Stern*, which was in print before the end of 1604.[7] His learned work on the star, *De Stella Nova* (Prague, 1606),[8] is much concerned with astrology: Kepler naturally regarded it as significant that the new star should have appeared at the time of a conjunction of Jupiter, Mars and Saturn and in a position close to that of the conjunction. The book also contains detailed observations of the position and brightness of the star: it being equally natural that Kepler should wish to establish that this new star, like Tycho's new star of 1572, was indeed a fixed star and not, for example, an atmospheric phenomenon. The brightness observations have enabled twentieth-century astronomers to identify Kepler's star as a Type I supernova, that is, a fairly old star (poor in elements other than hydrogen and helium) which exploded, the total energy involved in the explosion being roughly equal to the binding energy of the Sun (about 10^{49} ergs). The star would not have been bright enough to be visible to the naked eye before it exploded. The explanations Kepler considers are much less melodramatic than the modern one. However, one of them, the suggestion that the star had come closer to the Earth and then retreated again, leads him to discuss the much wider problem of the size of the Universe and the Sun's place in it.

Kepler establishes the special status of the Sun by an argument based on the observed appearance of the night sky (to the naked eye), the observed diameters of stars of various magnitudes and the assumption that stars of equal apparent magnitude are at equal distances from the Earth.[9] He considers the three second magnitude stars in Orion's belt. Two of them, S_1, S_2, are seen only 1°21′ apart on the sky. The angular diameter of the second star, S_2, seen from the Earth, T, is 2′. What size will S_2 appear when seen from S_1,

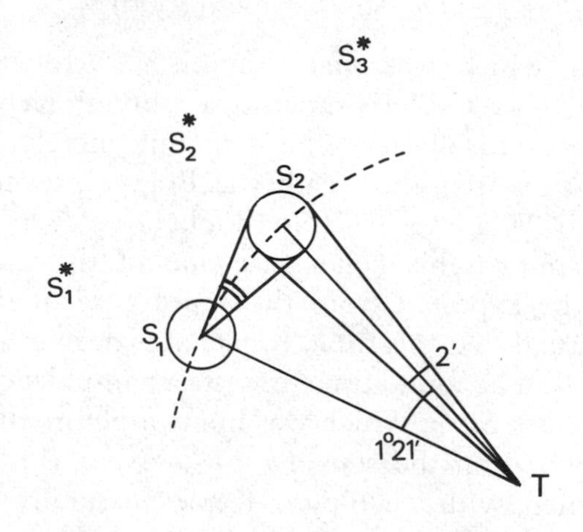

Figure 2.1 Stars in Orion's belt

assuming that the stars are equidistant from T? Kepler states that the answer is 'roughly' 2°45′, that is, more than five times the apparent diameter of the Sun seen from the Earth (which is about 30′). Neither Caspar nor Koyré appears to have noticed that the true answer, from the figures Kepler uses, should have been about 1°26′ (see the reworking of the calculation in Appendix 1). It seems, however, that Kepler himself did notice his mistake, since in the *Epitome Astronomiae Copernicanae* (in a passage where he still takes the diameter of a second magnitude star to be 2′) he says that the answer is roughly three times the apparent diameter of the Sun.[10] The mistake in the arithmetic does not, however, vitiate the rest of the argument since even with the lower value for the diameter it is still clear that the sky seen from S_1, S_2 or S_3 will be very unlike the sky seen from the Earth (or from the Sun) – a result which is incompatible with Kepler's assumption that an infinite Universe must be homogeneous, that is, that it must look the same irrespective of the position of the observer. Moreover, as Kepler points out, these three stars in Orion are by no means the closest stars of similar magnitudes that we observe in our sky. As will be seen from the reworking of Kepler's calculation, this result is independent of the distance of the stars from the Earth. Furthermore, it is clear that if, in pursuance of simplicity, the differences in apparent magnitudes of stars are ascribed entirely to their being at different distances from the Earth, then Kepler's argument extends to any grouping of stars whose magnitudes are similar to one another. Kepler's attempts to give a convincing value for the distance of the nearest fixed stars will be discussed in a later section of this chapter.

The assumption that stars of equal apparent magnitude are at the same distance from the Earth is not an unreasonable one, but Kepler seems to have been slightly unhappy with it, for he proceeds to show that his argument is still valid if the assumption is dropped. (Kepler's reasoning, which appears to be entirely correct, is discussed in Appendix 1.)

What vitiates Kepler's argument is neither his logic nor his mathematics but the observations he used. His value of 2′ for the diameter of a second magnitude star is, strangely enough, slightly larger than Tycho's value, as given in the *Progymnasmata*,[11] but the *true* value is zero and Kepler's calculation cannot in fact be carried out at all. Once telescopes were turned on the stars the diameters began to be questioned: in the *Sidereus Nuncius* (Venice, 1610) Galileo started the questioning, by pointing out that the stars do not seem to be magnified in the same proportion as the planets, and he spoke of the telescope's removing 'factitious rays'. By the time Kepler came to write Book I of the *Epitome* (Linz, 1618) it seems as though he may have had serious doubts about the apparent diameters of stars, for the argument which had covered about three sides in Chapter XXI of *De Stella Nova* has shrunk to a very short paragraph in the later work (with the arithmetical correction noted above). Finally, in Book IV of the *Epitome* (Linz, 1620) he wrote of the fixed stars that 'Experienced practitioners (*artifices*) say that telescopic observation shows no size or roundness of body [i.e. no disc]; rather, that the better the instrument the more the fixed stars show as pure points from which luminous rays come out and disperse themselves, in the manner of hair.'[12]

Epitome Astronomiae Copernicanae (1618 to 1621)

After *De Stella Nova* none of Kepler's works considers the problem of the size of the Universe until we come to the *Epitome*, a textbook written in dialogue form and published in three instalments: Books I to III appearing in 1618, Book IV in 1620 and Books V to VII in 1621.

Kepler deals with the question of the size of the Universe near the beginning of the *Epitome*, in the second part of Book I. To the question 'Do you think the centres of the stars lie in one spherical surface?' he gives the answer

> 'This is not certain. Since some are bright and some faint it is not unlikely that the faint ones seem faint because they are far away, high up in the aether, whereas those that seem bright are nearer to us. However, it is not absurd that two stars of unequal apparent magnitude should be at an equal distance from us.
>
> But it is certain that the planets are not in the same spherical surface as the fixed stars, but are beneath them, for they sometimes occult the fixed stars but are never occulted by them.'[13]

The *alter ego* then asks

> 'If nothing more certain than this is known about the fixed stars, it seems as though their domain were infinite: and this Sun of ours no more than one of the fixed stars, larger and brighter for us only because it is nearer than the fixed stars are ; and so round any of the fixed stars there might be a world such as that round us; Or which comes to the same thing, among the innumerable places in this infinite collection of fixed stars, this our World with its Sun will be only one, in no way different from the other places near other individual fixed stars: As shown at M in the adjoining figure?'[14] (See figure 2.2.)

The reply is that indeed such was the opinion of Bruno and of some of the ancients, but it does not follow that because the centres of the stars do not all lie in one spherical surface then their distribution in space must be uniform. Kepler suggests it is in fact as shown in the next diagram (see figure 2.3). The *alter ego* then naturally asks for arguments to support this contention.

Kepler's first argument is from packing: if the distribution of stars were uniform we should have only twelve nearest neighbours.

Figure 2.2 Homogeneous Universe (*Epitome* I, p. 35).

Our next nearest neighbours would be twice as far away and would be slightly more numerous, the next nearest would be three times as far away, and so on. This does not correspond to the appearance of the night sky, in which we see large numbers of stars of closely similar brightness and the gradation to fainter and fainter stars seems to proceed quite smoothly.

From the way in which Kepler presents this argument, one would suppose that the concept of packing was a thoroughly familiar one, except for the fact that he explains the number twelve as the number of vertices of a regular icosahedron. This explanation, together with the answer to the *alter ego*'s next question ('Why do you use the icosahedron?') raises more problems than it solves if one refers it to the problem in hand, which is equivalent to the problem of packing equal spheres. This problem does not in fact seem to have been a familiar one in Kepler's day. Indeed, historians credit Kepler himself with having given the first mathematical account of the problem, in

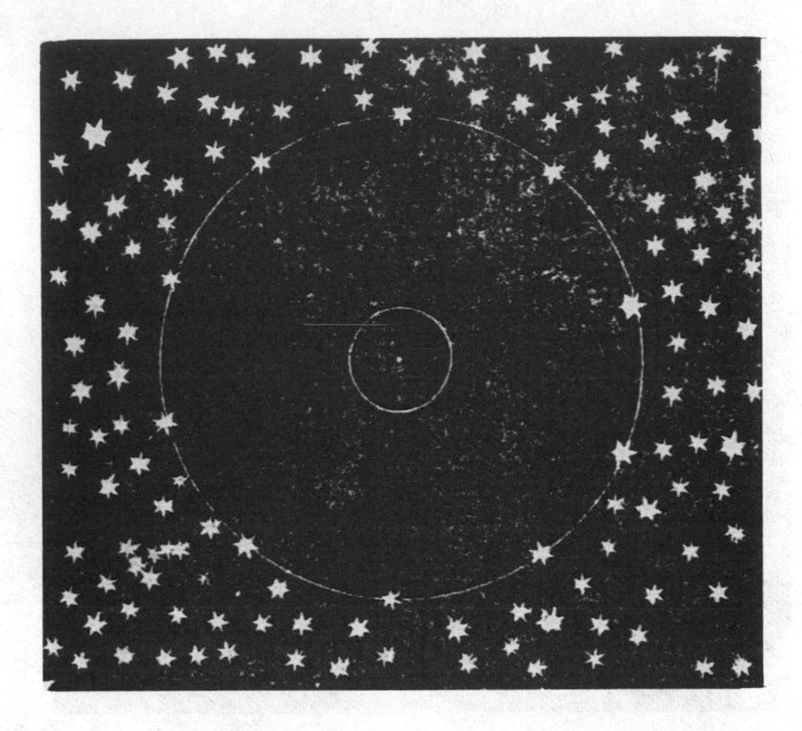

Figure 2.3 Non-homogeneous Universe (*Epitome* I, p. 36).

De Nive Sexangula (Prague, 1611),[15] a very short book, written as a New Year gift for Kepler's patron Wackher von Wackenfels, and despite its title concerned to a large extent with a detailed account of the structure of the honeycomb. In this work Kepler proceeds carefully step by step until he has described the packing of equal spheres which is equivalent to the packing of rhombic dodecahedra which defines the structure of the honeycomb. His account reads very much as if he expected his readers to be entirely unfamiliar with the idea of packing and this impression is confirmed by the fact that he has even gone so far as to include a pair of diagrams. One of the structures he asks the reader to imagine in the course of his exposition is a sphere packed round with twelve other spheres, all the spheres being of equal size. It is clear from the earlier stages of the procedure that the centres of the outer layer of spheres will lie at the twelve vertices of a cuboctahedron (see figure 2.4) (KGW *4*, pp. 268–9). The way Kepler describes the pattern is as three layers of spheres, the top and bottom layers having the centres of the spheres at the vertices of equilateral triangles and the middle layer having the centres of the spheres at the vertices of a regular hexagon and at its centre. One may think of the vertices and centre of a regular icosahedron as also lying in three layers, somewhat like those in the true arrangement, having equilateral triangles on the two outer layers, but with an inner layer which is not quite flat (see figure 2.5). (In figures 2.4 and 2.5 the centres in the outer layers have been indicated with rings, those in the inner ones with spots.) Kepler's explanation as to why he should refer to the icosahedron in this connection is that it is the regular polyhedron which best describes the distribution. It seems as though he felt that only a regular polyhedron was appropriate in explaining a fundamental property such as this.

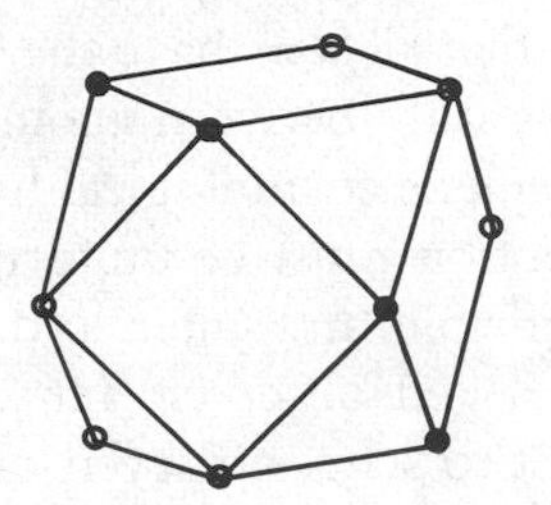

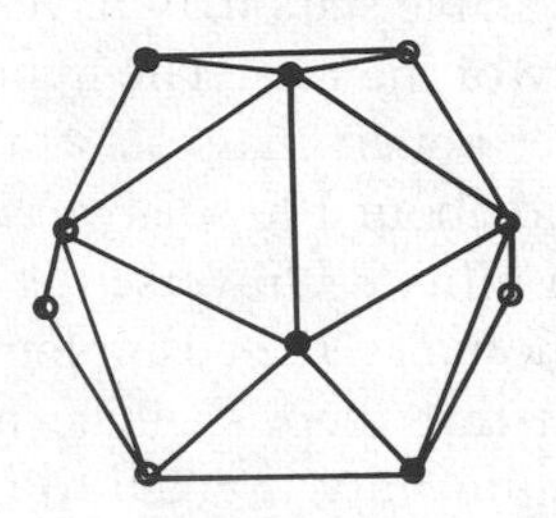

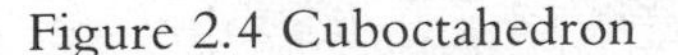

Figure 2.4 Cuboctahedron

Figure 2.5 Regular icosahedron

Having established that the distribution of stars cannot be uniform, Kepler briefly returns to the argument concerning the apparent diameters of stars which he used in *De Stella Nova*. However, he omits the part which considered the possibility that stars of the same magnitude might be at different distances and he gives no reference to his earlier and more detailed account. Kepler often does refer his reader to his other works so his silence in this instance may indicate that he was no longer satisfied with what he had written in *De Stella Nova*. On the other hand, he may merely have thought that his new argument from packing was more convincing than the old one. However, the conclusion he proceeds to draw, namely that the fixed stars are closer to one another than they are to the Sun, follows from the new argument only if their differences in apparent magnitude are ascribed almost entirely to their different distances from the Earth (or the Sun). Kepler accordingly goes on to discuss how a modification of this assumption would affect his argument. Uniform packing of identical stars implies that we should see only twelve very bright stars. Since we see many more, most of them must be further away, and intrinsically brighter than the nearer ones so as to appear equally bright to us. As Kepler points out, the region around the Sun would then be special in being surrounded by intrinsically fainter stars; and, moreover, the theory itself seems implausible. He concludes that 'it is more probable that stars of roughly equal apparent magnitude should be at roughly the same distance from us, and that the dense crowd of so many stars should form a sort of hollow sphere' (KGW 7, p. 44). In either case, however, the Sun's position would be special.

Kepler's astronomical hypothesis, that stars of equal apparent magnitude are at equal distances from us (that is, that all stars are of equal intrinsic luminosity) is a very reasonable one, and it was popular for a long time after Kepler's death. It has now been shown to be incorrect: stable stars may have luminosities up to about 10^7 times the luminosity of the Sun. The hypothesis that all stars have the same *mass* fares rather better: masses seem to vary only between about 0.4 solar masses and about 100 solar masses. Kepler's cosmological hypothesis, that in an infinite Universe the distribution must be uniform, is now usually known as 'the cosmological principle' and stated in the form of a demand that there shall be no privileged observer. However, the implied uniformity is taken to refer not to stars, nor even, usually, to galaxies, but to clusters of galaxies.

Kepler's argument from packing is, like his earlier argument from the apparent diameters of stars, logically and mathematically sound, but whereas the earlier argument was vitiated by the observational 'facts' it was based on, the new argument is vitiated by Kepler's very reasonable, but in fact erroneous, assumptions about the absolute luminosities of stars and the properties one should expect to characterise an infinite Universe.

The Milky Way

In both *De Stella Nova*[16] and in the *Epitome*[17] Kepler appeals to the appearance of the Milky Way in support of his description of the arrangement of the stars in space. In the later work he in fact proves that we are situated in the centre of the circle defined by the Milky Way, his method of proof being that used by Ptolemy to show that the Earth is in the centre of the equinoctial circle.[18] The Milky Way is then asserted to be so placed as to define the boundary between the Solar system and the region of the fixed stars.

The distance to the nearest fixed stars

The Copernican theory suggested a way of finding the distances of the fixed stars, namely by finding their annual parallax. Tycho Brahe naturally rose to the challenge of trying to observe such parallaxes. He was convinced that his observations were sufficiently accurate to allow him to detect any parallax greater than 1′ of arc and, since he did not detect any, he concluded that the distances to the stars must be so great that the observed stellar diameters of 2′ for a first magnitude star, 1′ 30″ for a second magnitude star, etc.,[19] would correspond to stellar radii comparable with the distance between the Earth and the Sun, if, that is, the Copernican theory were correct. He not unnaturally regarded this as a *reductio ad absurdum* disproof of the Copernican theory. In this connection, therefore, Tycho's observations were of no use at all to Kepler: the distances to the stars had to be decided on purely theoretical grounds.

The theory Kepler used is similar to that Archimedes ascribes to Aristarchus at the beginning of the *Sand-Reckoner*. He assumed that the ratio of the diameter of the body of the Sun to the maximum diameter of the orbit of Saturn should be equal to the ratio of the maximum diameter of the orbit of Saturn to the radius of the sphere of stars. Thus the orbit of Saturn would be a geometric mean between the sphere of the stars and the body of the Sun. Aristarchus, according to Archimedes, made the orbit of the Earth the geometric mean between the Sun and the stars.

There are two versions of Kepler's calculation, again in *De Stella Nova* and in the *Epitome*. They differ only in the numerical values ascribed to the distance between the Earth and the Sun (in each case taken as one tenth of the distance between the Sun and Saturn). In the earlier account this distance is taken as 1432 Earth radii and in the latter one as 3469 Earth radii. (The usual ancient value was 1200 Earth radii.) The resulting radii of the sphere of stars are $34,177,066\frac{2}{3}$[20] and 'roughly' 60,000,000 Earth radii.[21]

It is of interest to note that even Kepler's larger value for the distance between the Earth and the Sun is too small by a factor of about seven. His sizes for the Sun and planets would be too small by a similar factor and the Earth would therefore appear to him to be a more important component of the Solar system than it now appears to us. (For a wider discussion of this problem see Van Helden 1982.)

Kepler's finite cosmos

The part of Kepler's cosmology concerned with the structure of the Universe beyond the orbit of Saturn shows the same constancy as we shall find in the part of his cosmology that was concerned with the structure of the Solar system.

That Kepler was never forced to revise his beliefs about the distribution of the fixed stars is, in part, no more than a tribute to the impossibility of making any great appeal to observational evidence in constructing a cosmology in Kepler's day. But however strong Kepler's metaphysical revulsion from an infinite Universe may have been – and, as Koyré (1957) has shown, we have good reason to believe it was considerable, and firmly based on the work of Aristotle – he nevertheless felt the need to seek observational support for his model of the Universe. The first observational evidence to which he appealed, the apparent diameters of stars, was shown to be faulty by the improved observations made with the aid of telescopes, but Kepler managed to construct another argument based on observation, this time the relatively unassailable observation of the appearance of the night sky. However, in arguing from the appearance of the night sky Kepler was forced to make assumptions about the absolute luminosities of stars and the form of distribution to be expected in an infinite Universe. Both assumptions would have seemed very reasonable to the majority of Kepler's contemporaries. Both have since been shown to be erroneous. Yet if we remind ourselves that the 'observable Universe' of Kepler's day was in fact all but entirely made up of what are, by astronomical standards, very local objects – the bright stars of our local spiral arm, a few globular clusters in the galactic halo and only one extragalactic object, the Andromeda nebula (which looks like a sixth magnitude star) – we must acknowledge that the reality bears considerably more resemblance to the relatively compact model proposed by Kepler than it does to the infinite extent of stars described by Bruno.

Observational evidence was to destroy Kepler's model after his death but during his lifetime there were no convincing observational arguments against the model he had described in 1606:

> 'The perfection of the world is motion, which is as it were its life. For motion three things are required: mover, moved and place. The mover is the Sun, the moved are the planets Mercury to Saturn, the place is the outermost sphere, that of the fixed stars. If a physical thing may be expressed in mathematical terms, the

moved are a mean proportional between the mover and the place. . . . If the moved are in a physical sense a mean proportional between the mover and the place, it is the more likely that the same should be true mathematically, the diameter of the moved being a mean proportional between the diameter of the Sun, the mover, and the diameter of the place, the sphere of the fixed stars.'[22]

This model of the Universe seemed to Kepler to be so well established that he was apparently happy to explain it by analogy with the nature of the God who had created the Universe, the analogy being mentioned even in the *Epitome*, which is, by Kepler's standards, a very matter-of-fact work, as befits a textbook. Since Kepler believed that the Solar system had been shown to be a uniquely important part of God's Universe it was clearly in order for him to appeal to the very highest principles when searching for an explanation of its structure. As we shall see, his explanation was to make the structure of the Solar system a reflection of the properties of Euclidean space.

III
Mysterium Cosmographicum 1596

The *Mysterium Cosmographicum* (Tübingen, 1596), like all Kepler's later astronomical works, was written to the glory of God and dedicated to establishing what Kepler saw as the Copernican system. Any account of Kepler's work must therefore begin with Copernicus.

On the Revolution

Geocentric astronomy, based ultimately on the *Almagest*, enabled one to calculate the ratios of the radii of the various circles or spheres that were required to give a mathematical or physical explanation of the motion of any individual planet. It did not, however, provide any means of relating the resulting orbs or orbits one to another. The relationship between the orbits was provided by the physical assumption that there were no gaps between the spheres of neighbouring planets. It was further assumed that the sphere of the fixed stars lay immediately beyond the outermost planetary sphere. Moreover, geocentric astronomy also failed to provide any means of ascertaining the order in which the planetary spheres should be placed. However, there appears to have been almost perfect agreement that the order adopted should correspond to decreasing sidereal period (working inwards from the fixed stars). This was the order adopted by Ptolemy (*Almagest* IX, 1) – with the comment that it was not a matter of dispute among philosophers.

Copernicus knew that his own astronomy allowed him to find the actual sizes of the planetary spheres (and thus the order of the planets), that it enabled him, in fact, to construct a true system of planets. He makes this clear in the letter in which he dedicates *De Revolutionibus* to Pope Paul III: '. . . the orders and sizes of the stars, and all their orbs, and the heaven itself are so connected that nothing can be altered in any part of it without bringing confusion to all the remaining parts and to the whole universe'.[1] The order of the planets is described in *De Revolutionibus*, Book I, Ch. X (NKG, vol. II,

p. 25). This passage is prefaced by a discussion of all the orders that have been suggested by earlier astronomers, but Copernicus does not settle the matter, as he could have done, with the clear statement that his own system allows the order to be calculated. He merely justifies his adopted order by an appeal to the observed sidereal periods. The spacing of the planets is not mentioned at all in this chapter, and not only in the printed text of *De Revolutionibus* (see figure 3.1) but even in Copernicus' manuscript[2] we find that the diagram which accompanies this part of the text apparently shows a neat set of contiguous spheres. On closer inspection, however, it seems that in the printed diagram the spheres are in fact indicated as pure circles, without any thickness, except for the sphere of the Earth, which carries the sphere of the Moon.

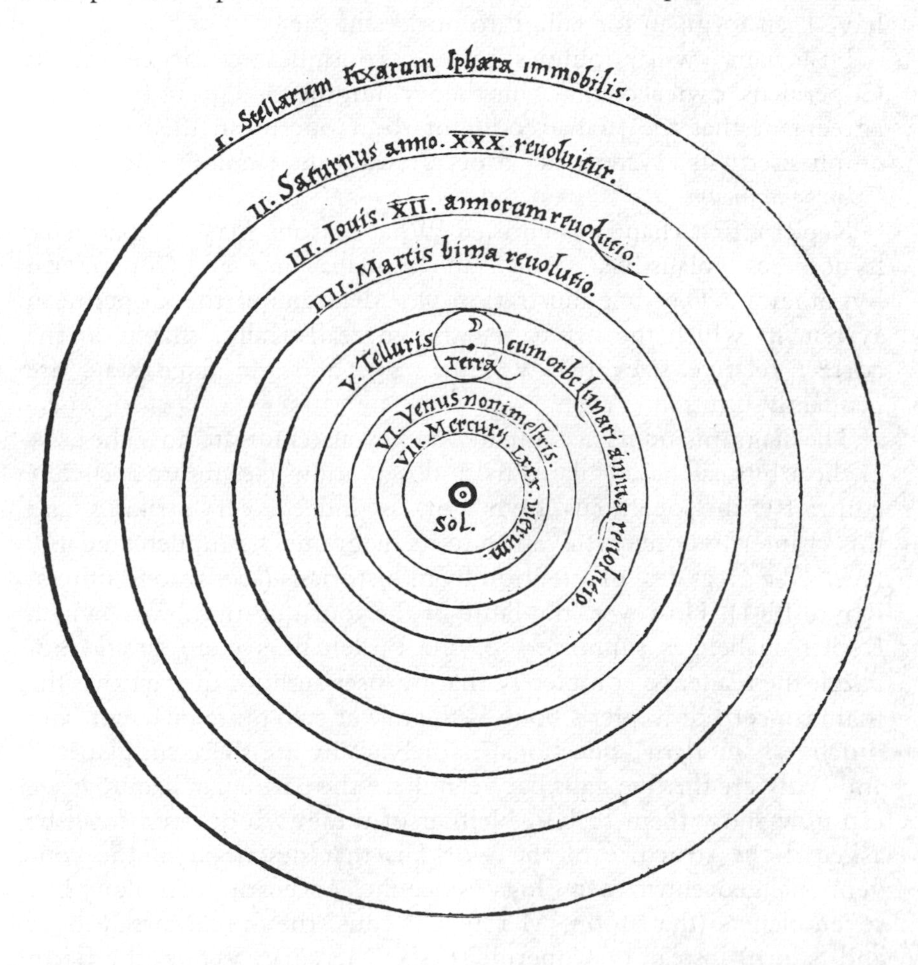

Figure 3.1 Planetary system (Copernicus *De Revolutionibus*, 1543, f9v)

The names must apply to the circle outside which each is written. (If the names were meant to apply to the annuli in which they appear then there would be a gap between the sphere of Mars and that of the Earth, though nowhere else.) Though the spheres in the diagram do turn out to be detached, the spaces shown between them are not at all realistic. Moreover, when the sizes of the individual planetary spheres are calculated, in *De Revolutionibus*, Book V, each planet is dealt with in a separate chapter, exactly in the manner of the *Almagest*. In fact, Copernicus' presentation is such that, if he had not actually stated in the dedicatory letter that his cosmology allowed him to give a systematic description of the planetary orbs, and to construct a unitary system including all the orbs, his readers might have been forgiven for failing to notice the fact.

Historians will doubtless long continue to argue about Copernicus' own account of his theory, but there seems to be general agreement that the first account of the Copernican theory which emphasised its systematic aspects was in Kepler's *Mysterium Cosmographicum*.

Kepler's first chapter is entitled 'What reasons make Copernicus' hypotheses plausible. And an explanation of Copernicus' hypotheses.' The first illustration is a diagram of the Copernican system in which the orbits are shown realistically, that is in the correct relative sizes (KGW *1*, p. 20), and the fixed stars are prudently omitted.

The diagram and its accompanying key also indicate how the sizes of the orbits can be calculated from observations (see figure 3.2). The contrast with Copernicus' own work is, indeed, very striking, and this chapter, whose text lives up to its diagrams, has understandably received a great deal of attention from historians (see, among others, Koyré 1961). However, the fame of the four questions with which Kepler challenges supporters of the Ptolemaic system should not dazzle the reader so completely that he loses sight of the fact that the main concern of Kepler's book is to answer two quite different, and much less 'modern', questions, namely 'Why are there six planets?' and 'Why are the five gaps between them the particular sizes that we can now show them to be?' Neither of these two questions can be asked if the structure of the world is that described in the conventional geocentric cosmology, since the geocentric cosmology had seven planets (the Moon, Mercury, Venus, the Sun, Mars, Jupiter and Saturn) instead of Copernicus' six (Mercury, Venus, the Earth with the Moon, Mars, Jupiter and Saturn). Moreover, as we have

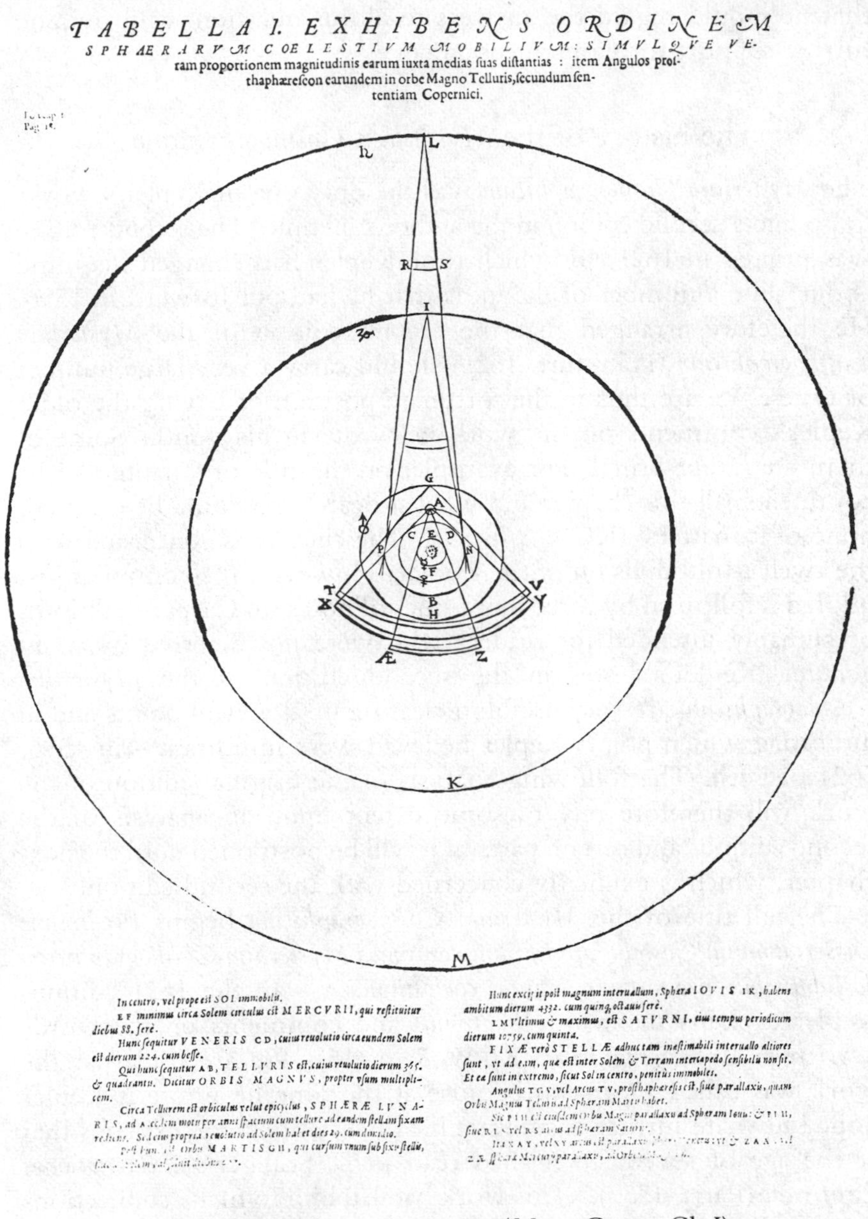

Figure 3.2 Planetary system (*Myst. Cosm.*, Ch I)

seen, the conventional cosmology did not admit a purely astronomical method of finding the distances between the planets. It was therefore reasonable for Kepler to begin his work with a chapter of arguments in favour of the Copernican system, and reasonable too that he should regard the answers to all his questions as providing further support for Copernicus' theory.

The history of the *Mysterium Cosmographicum*

The *Mysterium Cosmographicum* was the only one of Kepler's works to go into a second edition in the author's lifetime. The second edition was printed in 1621, by which time Kepler had changed his mind about quite a number of the ideas that he had put forward in 1596. He therefore arranged that the second edition of the *Mysterium Cosmographicum* (Frankfurt, 1621) should carry a very large number of notes. We are thus in the fortunate position of having the older Kepler's comments on the work he wrote in his youth. Some of them are rather brutal. For example, on the title of Chapter XI he comments 'all this chapter and what it deals with could be omitted, none of it matters' (KGW *8*, p. 62). The chapter is concerned with the twelve-fold division of the Zodiac. However, the comment just quoted is followed by three more sides of notes (in Caspar's edition), presumably intended for readers who were not deterred by it. In general, Kepler's notes in the second edition of the *Mysterium Cosmographicum* are very useful in clearing up doubtful points and in indicating which points Kepler believed were important – in 1596, 1621 or both. The following analysis of the original edition of the work will therefore rely to some extent upon an analysis of the second edition, and certain parts of it will be postponed until the next chapter, which is explicitly concerned with the second edition.

The full title of the *Mysterium Cosmographicum* begins *Prodromus Dissertationum Cosmographicarum, continens Mysterium Cosmographicum de admirabili proportione orbium coelestium*. . . . In the 1621 edition, Kepler explains the word *Prodromus* and comments on the words *Mysterium Cosmographicum* (KGW *8*, p. 15). We are told that the work was called *Prodromus* because at the time he wrote it Kepler hoped to write further cosmographic dissertations. He explains that at the age of seventeen he had read J. C. Scaliger's *Exercitationes Exotericae* (Paris, 1557).[3] This work had led him to make connections between his ideas on various different subjects 'such as the heavens, souls, spirits, elements, the nature of fire, the origin of springs, the ebb and flow of the sea, the shape of continental land masses, and of the seas between them, and similar things' (KGW *8*, p. 15, l. 8). So when he discovered the proportion between the celestial orbs he thought he might be able to find similar solutions to other problems. In the event, however, the 'prodromus' had not been followed by any 'epidromus', unless the reader might take the *Harmonice Mundi* (Linz, 1619) as its true successor, as being a much more thorough investigation of the same material.

Turning to the words *Mysterium Cosmographicum*, Kepler first complains that 'Cosmography' is sometimes used for 'Geography' and his work has on occasion been incorrectly catalogued by booksellers. Presumably he hopes that booksellers will read this note. On *Mysterium* he says 'I used *Mysterium* for *Arcanum* [i.e. secret, hidden] and I have offered my discovery as such, since I had never read anything like it in a book by any philosopher' (KGW *8*, p. 15, l. 33). Claims of complete originality are, in principle, always suspect – if only because accepting too many of them would make a history of science read like an account of a succession of miracles. However, it seems reasonable to accept Kepler's claim in this case: his disposition seems to have been to give too much credit, rather than too little, to the people from whom he had learned something he considered valuable – witness his attitude to Copernicus – and he gives a fairly full account of how he came to his theory. The problem will be considered in more detail below, in connection with this description of the theory's origin.

Kepler's annoyance at finding his work mis-classified by booksellers accounts for his wishing to explain the word *cosmographicum*. There is, however, no comparably simple reason why he should have wished to explain the word *mysterium*. Rosen (1975) has discussed at some length the various cuts and alterations that were made in the *Mysterium Cosmographicum* at the order of the University of Tübingen, under whose auspices the work was published, but it does not appear that there was any discussion of the title of the work.

The system of planets and polyhedra described in the *Mysterium Cosmographicum*

The Copernican description of the planetary system makes the circles which formed the deferents of the superior planets in the geocentric system into the true orbits of the planets. The circles which were the epicycles are then seen as no more than reflections of the annual orbital motion of the Earth. For the inferior planets, the former deferents are the orbit of the Earth and the former epicycles the true orbits. Since the ratios of the radii of the epicycles to the radii of the deferents can be calculated from observations, we can go on to calculate the true dimensions of the orbits of the planets, all dimensions being measured in terms of the radius of the orbit of the Earth. This calculation is not by any means as simple as the above brief account would suggest, since the heliostatic orbits, even when calculated from the observational data available to Copernicus, are not simple circles. However, a calculation can be carried out along the lines just indicated, and it shows that the orbits of the planets are widely separated.

It is important at this point to note that while Copernicus' planetary theory does not involve noticeably fewer circles than the geocentric theory it was proposed to replace – indeed, some methods of counting show Copernicus' theory as involving more circles than its predecessor – the Copernican orbits are very much closer to being circles than the geocentric orbits were, and the small Copernican epicycles make only a tiny contribution to filling the gaps between the circles which mark the mean orbits of the planets.

Since Kepler asks supporters of the geocentric system to explain the sizes of their epicycles[4] it is hardly a matter for surprise that he should have taken it upon himself to explain the apparently arbitrary gaps between the planetary orbits in the Copernican system. However, the fact that it was the sizes of the gaps, rather than the sizes of the orbits, that he chose to attempt to explain suggests that he was still thinking at least partly in terms of the nested system of spheres shown in the geocentric cosmologies. It was only later that he was to concentrate his attention upon the Sun-planet relationship: when, that is, he was faced with an astronomical rather than a cosmological task, namely to find the orbit of Mars in order to be able to predict future positions of the planet.

The cosmological theory described in the *Mysterium Cosmographicum* explains the gaps between neighbouring planetary orbits by the relations between the circumspheres and inspheres of the five Platonic solids (the five convex regular polyhedra, shown in figure 3.3). The mathematical basis of the theory is derived from Book XIII of Euclid's *Elements*. In modern terms it may be expressed as follows: The faces of a regular polyhedron are all regular polygons of the same shape and they meet in the same way at every vertex of the solid. Therefore, by symmetry, all the vertices must lie on a sphere and all the centres of the faces must lie on another sphere, which will touch the faces at these points. Again by symmetry, the two spheres will be concentric. They are known as the circumsphere and the insphere of the polyhedron, and they are analogous to the circumcircle and incircle of a regular polygon. The ratio of the radii of the two spheres for any given regular polyhedron is a fixed quantity and can be calculated fairly easily (even in the most difficult case). Appendix 2 shows a possible method of carrying out the calculation for the fairly simple case of the cube. In fact, the five Platonic solids yield only three ratios, since it can be shown that the ratio of the spheres for the cube is equal to that for the octahedron, and the ratio for the dodecahedron is equal to that for the icosahedron. The ratios are shown in table 3.1.

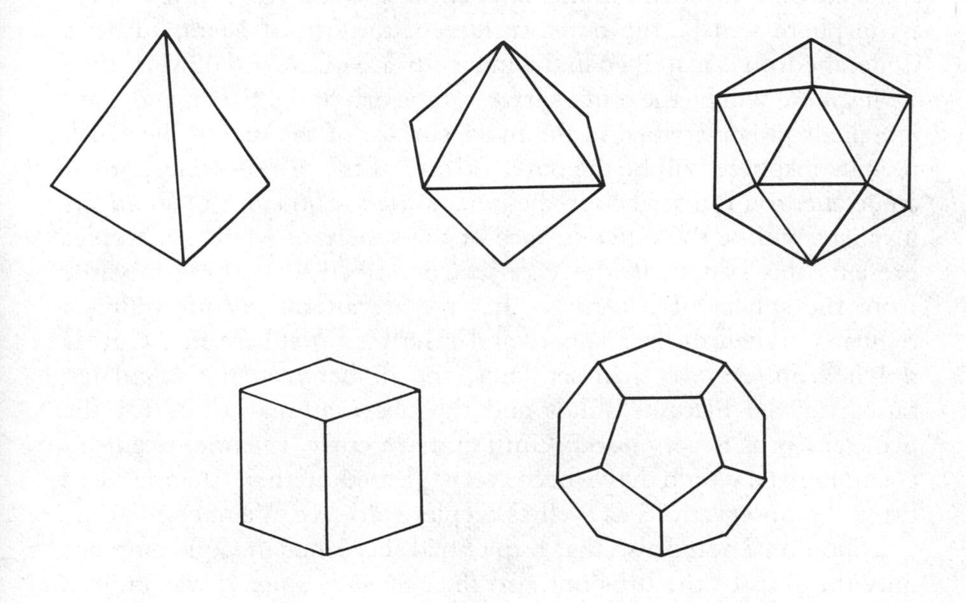

Figure 3.3 The five convex regular polyhedra

Table 3.1 *Radii of circumspheres and inspheres of the Platonic solids*

solid	R_c:R_i	R_c/R_i (to 5 fig.)
tetrahedron	$3:1$	3
cube octahedron	$\sqrt{3}:1$	1.7321
dodecahedron icosahedron	$\sqrt{15-6\sqrt{5}}:1$	1.2584

From reading the text of the *Mysterium Cosmographicum*, one has the impression that Kepler takes the inner radius of each planetary orb as being equal to the planet's minimum distance from the centre of the orbit of the Earth and the outer radius as being equal to its maximum distance from it. However, Aiton (1977, 1981), has checked Kepler's results and finds that the radii of the surfaces of the orb of the Earth have in fact been measured from the Sun, giving the orb a thickness equal to the eccentricity of the orbit of the Earth. It turns out that the dimensions of these planetary orbs are such that if a cube is inscribed in the inner surface of the orb of Saturn then its insphere will be the outer surface of the orb of Jupiter, and if a tetrahedron is inscribed in the inner surface of the orb of Jupiter then its insphere will be the outer surface of the orb of Mars, and if a dodecahedron is inscribed in the inner surface of the orb of Mars then its insphere will be the outer surface of the orb of the Earth, and if an icosahedron is inscribed in the inner surface of the orb of the Earth then its insphere will be the outer surface of the orb of Venus, and if an octahedron is inscribed in the inner surface of the orb of Venus its insphere will be the outer surface of the sphere of Mercury. Kepler presents this system in the way just described – working inwards from the sphere of Saturn so that we are alternately inscribing a regular polyhedron in a sphere and inscribing a sphere in a regular polyhedron. As we shall see later, the fit between the calculated ratios for the Platonic solids and the observational ratios for the planetary orbs is very good. Until quite recently, twentieth-century cosmologists would have been very pleased if their theories had fitted the observations as well as Kepler's do (see Weinberg 1977).

It should be noted also that Kepler had succeeded in explaining not only the sizes of the orbs but also their *number*, since it was known that there were only five convex regular polyhedra. (A proof of this fact is given as a scholium to Proposition 18 of Book XIII of the *Elements* and Kepler gives a summary of it in the *Mysterium Cosmographicum*, together with a reference to Euclid. Since the proof was, clearly, well known to Kepler's readers his repetition of it presumably indicates that he wished to stress its relevance to his theory.)

The brief account of Kepler's theory in Chapter II of the *Mysterium Cosmographicum* is illustrated with a very handsome engraved plate which folds out of the book (see figure 3.4). Some disembodied dotted lines have been drawn to indicate the positions of the centres of faces of the polyhedra and the points where their vertices touch the

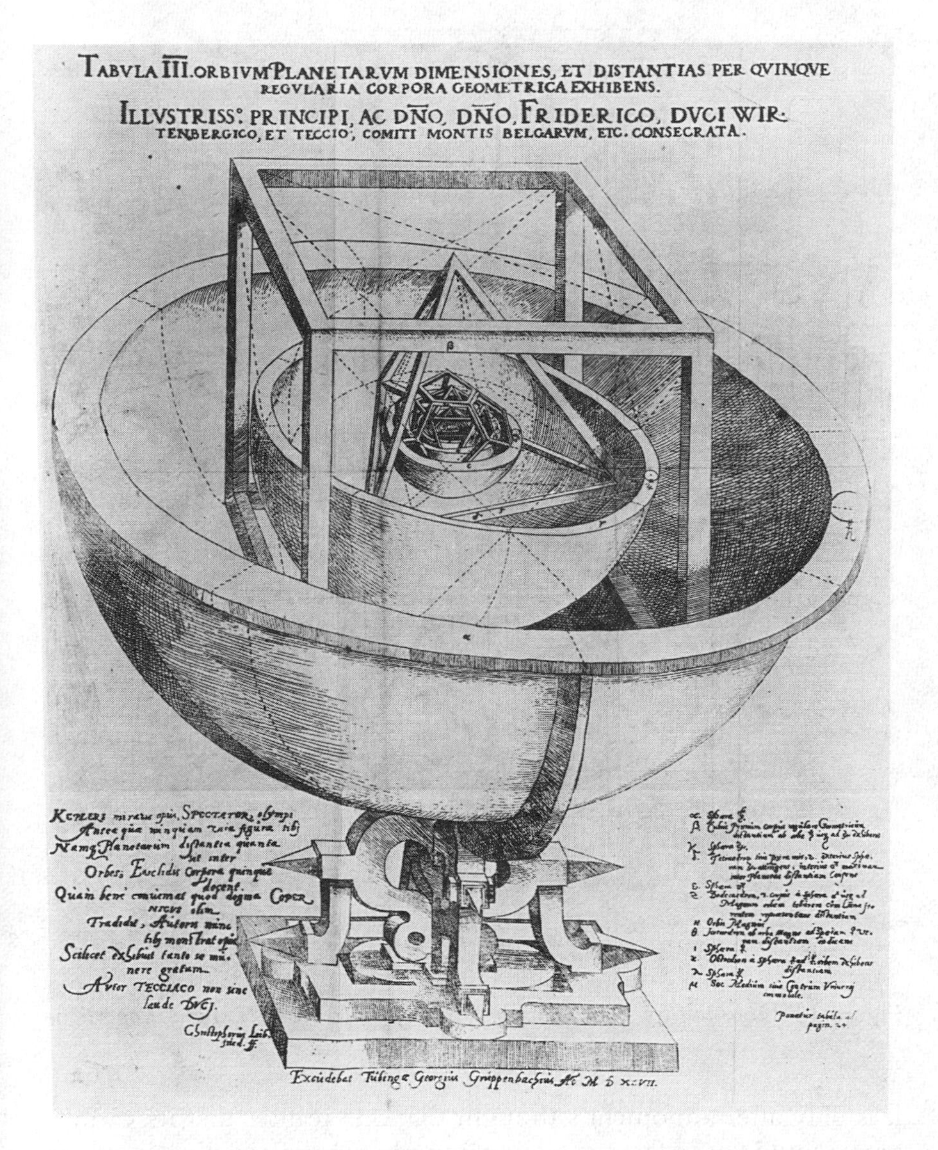

Figure 3.4 Orbs and polyhedra (*Mysterium Cosmographicum*, 1596, Ch. II)

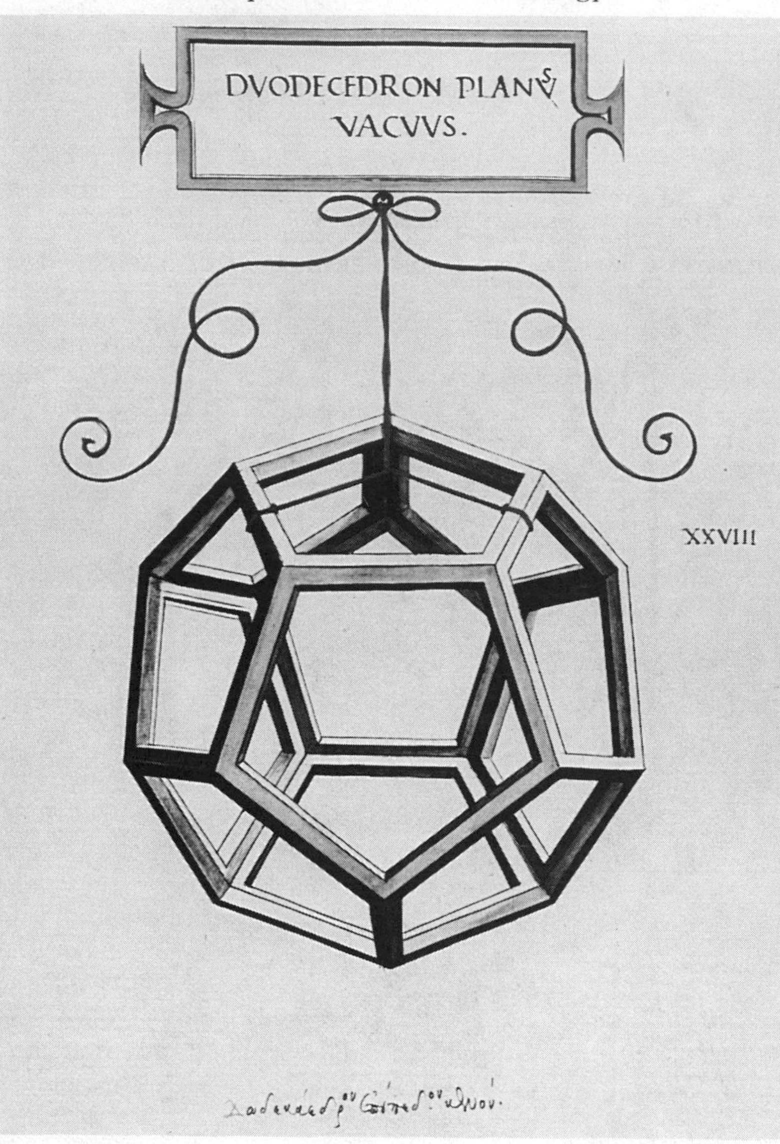

Figure 3.5 Drawing by Leonardo da Vinci for Pacioli *De Divina Proportione* (Venice, 1509)

it is only after a moment's thought that one realises that its various parts would not in fact support one another in the positions in which they are shown. The illusionistic rendering of impossible objects was, of course, no novelty. Beautiful examples are provided by the drawings Leonardo da Vinci (1452–1519) made to illustrate Luca Pacioli's *De Divina Proportione* (Venice, 1509) (see figure 3.5). (The published wood engravings based on the drawings have lost the strange intensity of the originals. Perhaps the engraver rashly removed slight distortions Leonardo had used to enhance the impression of a third dimension?)

It is possible that Kepler's illustrator made use of the published versions of Leonardo's pictures. In any case, the polyhedra shown in Kepler's plate are, like Leonardo's, shown as having a certain thickness. In Kepler's theory, however, the polyhedra were used as purely mathematical entities. So, too, were the spheres, though Tycho Brahe's reading of the *Mysterium Cosmographicum* seems to have left him with the impression that Kepler did believe in solid celestial spheres. There is certainly nothing in Kepler's theory that would tend to disprove their existence, but we do know that Kepler did not, even at this early stage, believe in solid spheres: in Chapter XVI of the *Mysterium Cosmographicum* he calls the idea of such spheres 'absurd' and 'monstrous'.[5] In the plate which accompanies Chapter II (figure 3.4), the thickness of each sphere has apparently been explained by showing it to contain a small circle which carries the planet. The circles shown are not, in fact, epicycles, since the deferent would not be concentric with the planetary sphere. The weakness of the plate as an astronomical diagram is underlined by the brevity of the key, which contains only twelve entries. In any case, the uncooperative nature of the astronomical facts ensures that the inner part of the picture is exceedingly difficult to read.

A very much clearer, and drier, diagram is provided to accompany the more detailed account of the theory in Chapter XIV (KGW *1*, p. 49) (figure 3.6). In this diagram, the rather vague indication that the Sun is 'the middle or centre of the Universe and at rest' is replaced by the more precise note that the centre of the Great Orb is the centre of all the spheres 'and close to the body of the Sun itself'. The next chapter contains two diagrams of the region close to the Sun, showing the centres of the deferents of planets, at the time of Ptolemy and at the time of Copernicus (KGW *1*, pp. 52, 53).

This kind of detail is, of course, required for any serious attempt to fit the cosmological theory to the astronomical facts deduced from observations with the help of the Copernican theory. It is presumably this connection with observations which lies behind a feature of the diagram in Chapter XIV which has no counterpart in the elaborate engraved plate which accompanies Chapter II, namely the circle of the Zodiac which appears round the outside of the diagram. Between this circle and the orb of Saturn is a narrow space

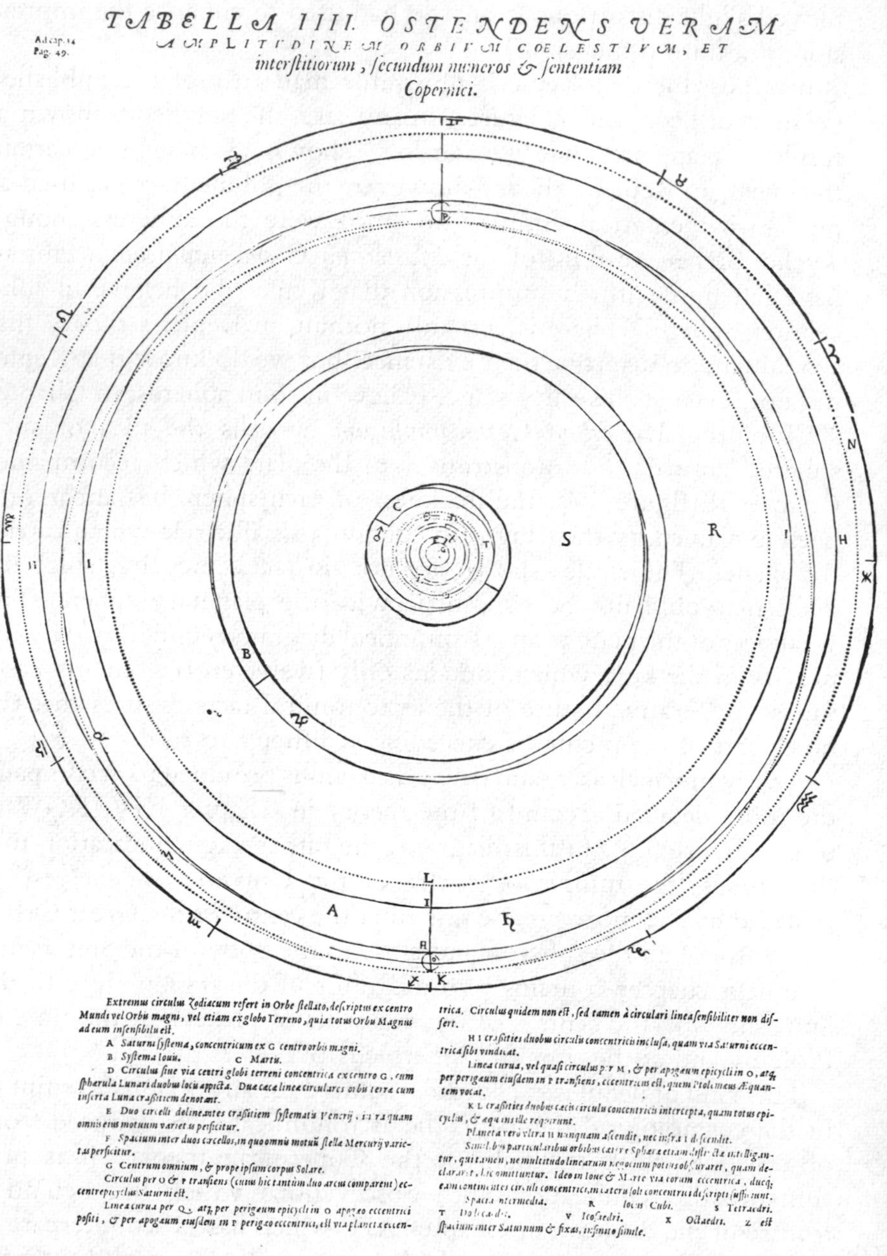

Ad cap. 14 Pag. 49.

TABELLA IIII. OSTENDENS VERAM AMPLITUDINEM ORBIUM COELESTIUM, ET *interstitiorum, secundum numeros & sententiam Copernici.*

Extremus circulus Zodiacum refert in Orbe stellato, descriptus ex centro Mundi vel Orbis magni, vel etiam ex globo Terreno, quia totus Orbis Magnus ad eum insensibilis est.

A *Saturni systema, concentricum ex* G *centro orbis magni.*

B *Systema Iouis.* C *Martis.*

D *Circulus siue via centri globi terreni concentrica ex centro* G, *cum sphærula Lunari duobus locis appicta. Duæ cacæ lineæ circulares orbis terræ cum inserta Luna crassitiem denotant.*

E *Duo circelli delineantes crassitiem systematis Veneris, intra quam omnis eius motuum varietas perficitur.*

F *Spacium inter duos circellos, in quo omnis motuũ stellæ Mercurij varietas perficitur.*

G *Centrum omnium, & prope ipsum corpus Solare.*

Circulus per O *&* P *transiens (cuius hic tantùm duo arcus comparent) eccentrepicyclus Saturni est.*

Linea curua per Q, *atq; per perigæum epicycli in* O *apogæo eccentrici positi, & per apogæum eiusdem in* P *perigæo eccentrici, est via planetæ eccentrica. Circulus quidem non est, sed tamen à circulari linea sensibiliter non differt.*

H I *crassities duobus circulis concentricis inclusa, quam via Saturni eccentrica sibi vindicat.*

Linea curua, vel quasi circulus per M, *& per apogæum epicycli in* O, *atq; per perigæum eiusdem in* P *transiens, eccentricus est, quem Ptolemæus Æquantem vocat.*

K L *crassities duobus cæcis circulis concentricis intercepta, quam totus epicyclus, & æquans ille requirunt.*

Planeta verò vltra H *nunquam ascendit, nec infra* I *descendit.*

Similibus particularibus orbibus cæteræ Sphæræ etiam distinctæ intelliguntur, qui tamen, ne multitudo linearum negocium potius obscuraret, quam declararet, hic omittuntur. Ideo in Ioue & Marte via eorum eccentrica, duoq; eam continentes circuli concentrici, in cæteris soli concentrici descripti sufficiunt.

Spacia intermedia. R *locus Cubi.* S *Tetraedri.* T *Dodecaedri.* V *Icosaedri.* X *Octaedri.* Z *est spacium inter Saturnum & fixas, infinito simile.*

Figure 3.6 Planetary orbs (*Myst. Cosm.*, Ch. XIV)

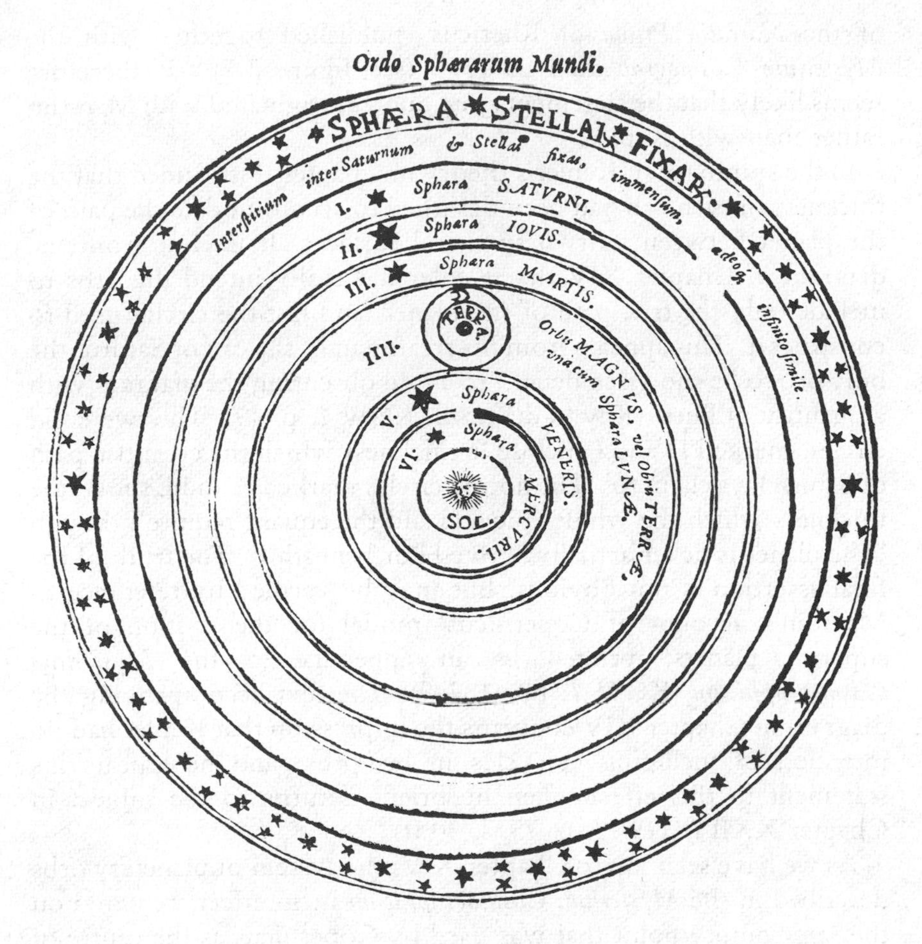

Figure 3.7 Planetary orbs (Rheticus *Narratio Prima*, ed. Maestlin, 1596, p. 117)

marked Z, explained in the key 'Z is the space between Saturn and the fixed stars, like infinity'. The first line of the key reads 'The outermost circle, with the Zodiac, indicates the orb of the fixed stars, whose centre is the centre of the World or that of the Great Orb, or even the body of the Earth, because the whole of the Great Orb is insensibly small in comparison with it.' These two statements seem to constitute the only mention of the relationship of the planetary system to the fixed stars in the *Mysterium Cosmographicum*. They echo both Copernicus and, more closely, the wording in the diagram of the Universe which, it appears, Maestlin contributed to his edition of the *Narratio Prima* of Rheticus, published together with the *Mysterium Cosmographicum* in 1596 (see figure 3.7).[6] It therefore seems likely that the statements just quoted originated with Maestlin rather than with Kepler.

In the summary of Kepler's theory in Chapter II it seemed that the thickness of each orb had been designed to accommodate the path of the planet between its two spherical surfaces. It is clear from the diagram in Chapter XIV that Kepler did really intend the orbs to include only the true path of the planet, and not the circles used to construct it. This appears from his treatment of the orb of Saturn, the only orb to be shown in detail, 'to avoid obscuring the diagram with a quantity of lines' (Key to diagram, KGW *1*, p. 49). The two solid circles, marked H and I enclose 'a thickness which the eccentric path of Saturn lays claim to'. The dotted circles marked K and L show 'the thickness which the whole epicycle and the equant require', though 'The planet is never actually above H or beneath I'. The truth of the final assertion is not obvious, but may be verified by reference to Maestlin's account of Copernicus' model for the motion of the superior planets, printed in an appendix to the *Mysterium Cosmographicum* (KGW *1*, pp, 137–9). The text accompanying the diagram in Chapter XIV confirms the impression that Kepler had no intention of including epicycles in his orbs, and he repeats his statement to this effect when he briefly returns to the subject in Chapter XXII (KGW *1*, p. 75, l. 30 ff).

As we have seen, up to Chapter XIV, the system of planetary orbs described in the *Mysterium Cosmographicum* is, in effect, centered on the same empty point that was used by Copernicus as the centre of his system, namely the centre of the Great Orb (though the radii defining the orb of the Earth have been measured from the Sun). However, Kepler is concerned not with the circles used to construct the orbits but with the actual paths of the planets, so there is no compelling mathematical reason for choosing this point. In the next chapter, Chapter XV, the centre of the system is transferred to the Sun. Kepler gives no justification for this. Instead, he justifies Copernicus' original choice of centre, which, he suggests, was made to shorten his calculations and to avoid upsetting readers by too great a departure from Ptolemy (KGW *1*, p. 50, ll. 31–4). *Mutatis mutandis* this may well reflect Kepler's own motives in originally adopting Copernicus' centre. His own choice was henceforth to be the Sun.

The origin of Kepler's theory

In the preface to the *Mysterium Cosmographicum* Kepler describes how he was led to formulate his theory.

He recounts how he studied astronomy at Tübingen under Maestlin and then, while teaching mathematics and astronomy at Graz, found that astronomical studies were absorbing more and more of his attention, whereas at Tübingen he had given much of his time to the study of theology. It seems that Kepler, like many another, found that having to teach a subject made him learn a great deal about it. The task which he had set himself was 'to give physical or, if you prefer, metaphysical reasons' for what Copernicus had described in mathematical terms (KGW *1*, p. 8, l. 17). The reference to Copernicus' work is rather vague, but Kepler presumably means that he will, for instance, give reasons for the spacing of the planets. If he had meant that he would give reasons for accepting the Copernican System he surely would not have characterised Copernicus' reasoning as merely mathematical. It should, however, be noted that Kepler's works are a rich mine of quotations which in other circumstances would suffice to convince any reasonable historian that the writer had not read *De Revolutionibus* with any great attention. Nevertheless, as Aiton (1977, 1981) has pointed out, we know that while he was writing the *Mysterium Cosmographicum* Kepler used not only Rheticus' *Narratio Prima* but also *De Revolutionibus* itself. In a letter dated 3 October 1595 he had mentioned to Maestlin that he now possessed his own copy of Copernicus' work (KGW *13*, letter 23, l. 455, p. 45). However, even in the *Mysterium Cosmographicum*, the work in which Kepler's Copernicanism most closely resembles that of Copernicus, it was not Kepler's purpose to give an historically correct account of the work of Copernicus. There were three things, he tells us, that he particularly wanted to explain: 'the Number, Size and Motions of the Orbs' (KGW *1*, p. 9, 1. 33). All three of these were different in the Copernican system from what they had been in the conventional geocentric system: there were six planetary orbs, instead of seven, their sizes could be calculated by astronomical methods and the true motions were very much more nearly circular. Kepler adds

> 'I was made bold to attempt this by the beautiful harmony that exists between the parts that are at rest, the Sun, the fixed stars and the intermediate space, and God the Father, the Son and the Holy Ghost: a similarity I shall pursue further through Cosmography.

Since the parts that are at rest are disposed in this way, I did not doubt that the moving parts would also be harmonious' (KGW *1*, p. 9, 1.37 ff).

Kepler first describes his false starts, which included an attempt to give an explanation of the sizes of the orbs in terms of the properties of numbers,[7] and an attempt to describe the motions by reference to the lengths of lines cut off between one side of a square and a quadrant of a circle drawn with its centre at one vertex of the square and the side of the square as its radius (see figure 3.8). This second idea somewhat resembles Galileo's attempt, in the *Dialogo*, to explain the orbital velocities of the planets by regarding them as having fallen from rest through distancs related to the dimensions of their orbits. We know that Galileo derived the numerical data for this theory from the *Mysterium Cosmographicum* (see Drake 1978, pp. 154–5, and 63–5).

The idea which was to lead Kepler to explain the planetary orbs by means of the Platonic solids came to him quite suddenly on 19 July 1595 (New Style, as are all dates after 1582 unless otherwise

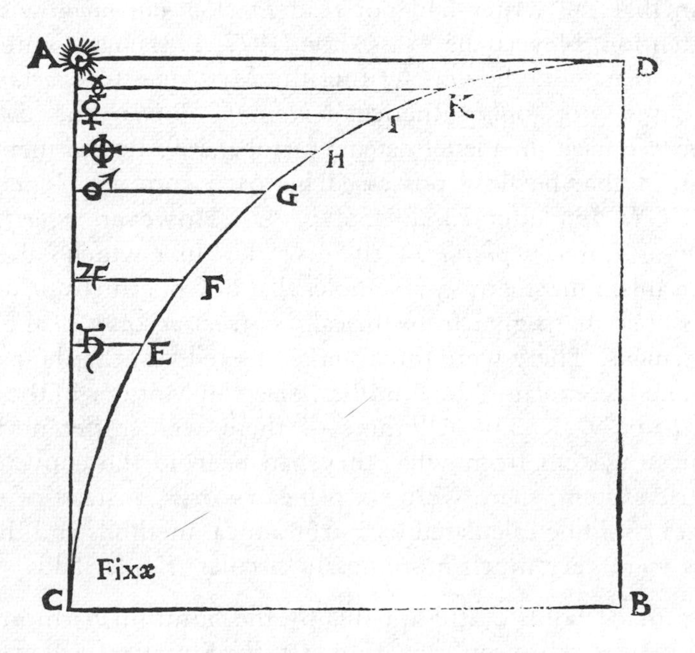

Figure 3.8 Motions of planets (*Myst. Cosm.*, Preface)

indicated). Characteristically, Kepler interpreted the slightness of the occasion for this discovery as indicating that God had answered his prayer for guidance.[8]

> 'So on 9 or 19 July of the year of 1595, I was about to show my students the way Great Conjunctions jump eight signs at a time and how they progress step by step from one Trigon into another; I had drawn many triangles, or quasi-triangles, in the same circle, in such a way that the end of one was the beginning of the next. So the points where the sides of the triangles cut one another indicated a smaller circle. For the radius of a circle inscribed in a triangle [i.e. an equilateral triangle] is half the radius of the circumscribed circle'[9] (see figure 3.9).

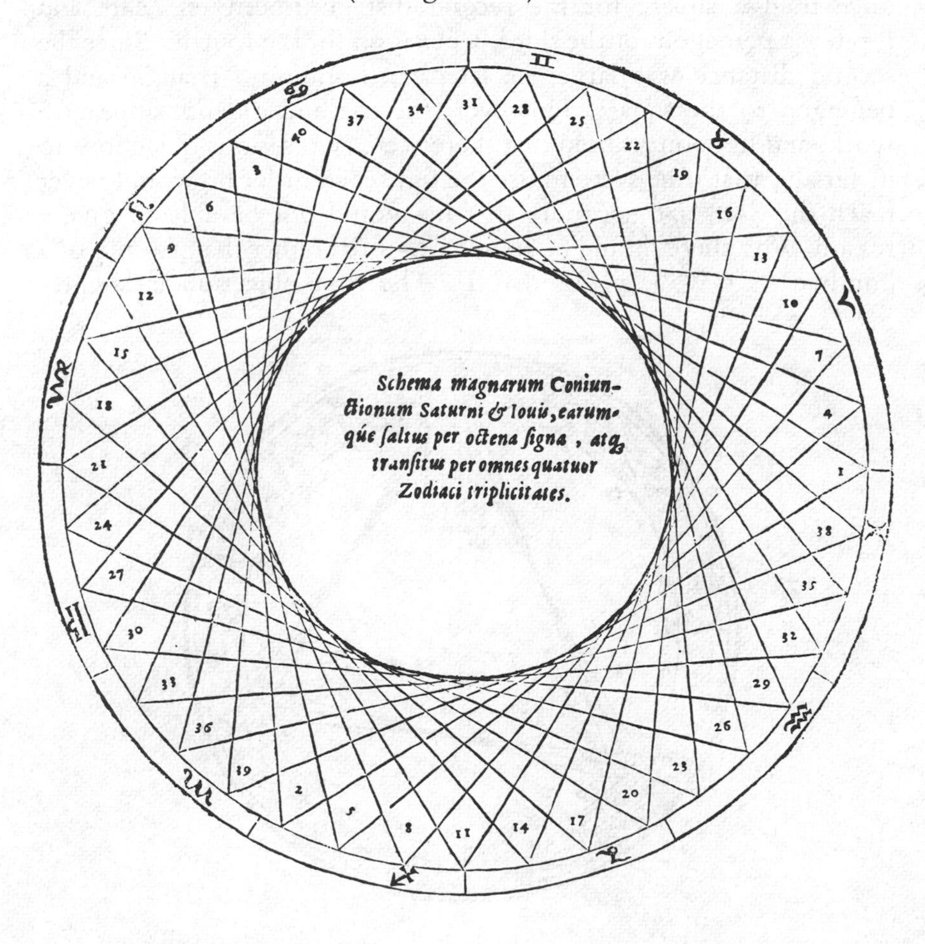

Figure 3.9 Great conjunctions (*Myst. Cosm.*, Preface)

Kepler's diagram contains so many 'quasi-triangles' that the trigons are not easy to pick out. Figure 3.10 shows a diagram which indicates one trigon, the fiery trigon, reasonably clearly, together with the dates at which Great Conjunctions occurred in its three signs. This diagram accompanies a chapter of *De Stella Nova* (Prague, 1606) that deals with the names of the four trigons and pronounces them to be derived from human choice rather than from Nature.[10]

Having noticed the smaller circle, Kepler thought that the proportion between the two circles looked similar to the proportion between the orbs of Saturn and Jupiter. He also noted that the triangle was the first of the regular polygons just as Saturn and Jupiter were the first of the planets. He describes the next stage: 'I at once tried a square for the second distance, between Mars and Jupiter, a pentagon for the third, a hexagon for the fourth.' Since the second distance was large, he later tried adding a triangle and a pentagon to the square. However, the method did not appear to work, and he soon realised that there were two serious objections to it. Firstly, that if he were to use the figures in order he would never reach the Sun, and secondly that he would not be able to give a reason 'Why there should be six moving orbs rather than twenty or a hundred' (KGW *1,* p. 12, 1.10ff). The first objection is slightly

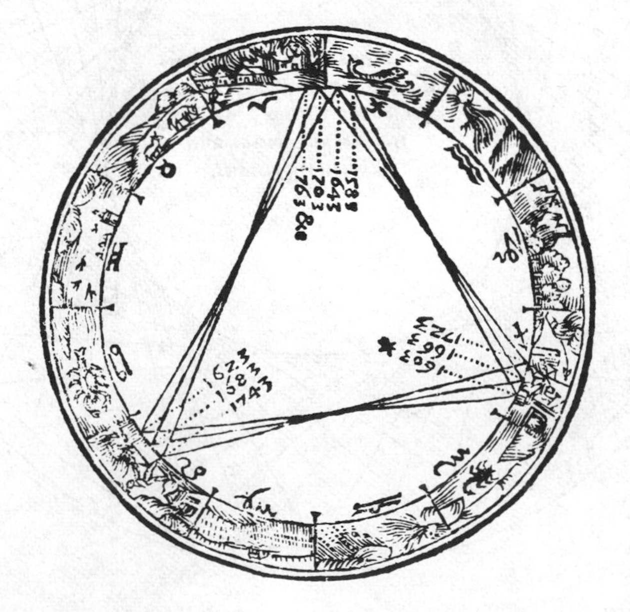

Figure 3.10 Great conjunctions (*De Stella Nova*, Ch. VI, p. 25)

strange in that even a succession of triangles, each one halving the radius for the next sphere, will never reach to the centre. (The objection does, however, suggest that at this stage Kepler may have been looking for a theory which would relate the size of the orb of Mercury to the size of the Sun, something which his final theory does not do.) Presumably Kepler was not thinking of the problem of reaching exactly to the centre but merely noticing that as the number of sides of the polygon increases so the circumcircle and the incircle become closer and closer in size – a fact which Archimedes had used to find the circumference of a circle and hence the value of π. (Appendix 3 shows something of the geometrical basis of Kepler's theory.)

Kepler was not entirely discouraged by his failure. His description of it concludes with the note 'the figures pleased me, as being quantities, and things prior to the heavens. For quantity was created at first, with body, the heavens were created the next day.' This is quite clearly a reference to the first chapter of Genesis: 'In the beginning God created the heaven and the earth. . . . And the evening and the morning were the first day. . . . And God called the firmament Heaven. And the evening and the morning were the second day.' Kepler seems never to have found any conflict between his religious convictions and his natural philosophy. His commonsense account of the status of the Bible in regard to matters of natural philosophy was eventually printed in the preface to the *Astronomia Nova* (Heidelberg, 1609), but it had been written for the *Mysterium Cosmographicum*, and suppressed in deference to the sensibilities of the University of Tübingen (see Rosen, 1975). When Kepler does refer to the Creation it is usually in terms that are derived from Plato rather than from the Bible – as we have seen, Kepler regarded *Timaeus* as a commentary on *Genesis*. For example, in the *Harmonice Mundi* (Linz, 1619) he says '. . . our Faith holds that the World, which had no previous existence, was created by God in weight, measure and number, that is in accordance with ideas coeternal with Him'.[11]

According to Kepler's account, it was his attempt to select five particular polygons to describe the six Copernican orbs that led him to think about polyhedra instead of polygons. Why should there be plane figures between the three-dimensional orbs? Five particular *solid* figures were easily distinguishable from all others, namely the five Platonic solids, described in Book XIII of the *Elements*: 'See, Reader, this discovery and the matter of all this little work'.[12] Kepler further points out that the scholium to Proposition 18 of Book XIII of the *Elements* proves that it is not possible to construct any more than five regular bodies. His theory was thus established on a secure, Euclidean, mathematical base. It was, however, a base which Kepler himself was to undermine, by discovering two new regular polyhedra (which he recognised as such). It appears that he had discovered at least one of these new polyhedra by 1599 (see Field 1979a), but his first full description of them appeared only in 1619, in Book II of the *Harmonice Mundi*. Discussion of them will accordingly be postponed until Chapter V below.

The second edition of the *Mysterium Cosmographicum* reprints the passage in which Kepler describes the origin of his theory without the addition of any note that might be taken as modifying this account. Moreover, as we have seen, Kepler's note on the title of the *Mysterium Cosmographicum* makes an explicit claim that the theory was original, describing it as *inventum*, as it is described also in the passage from the preface quoted in the last paragraph 'See, Reader, this discovery (*inventum*) and the matter of all this little work'. Maestlin uses the same word in referring to the theory in several letters to officials at Tübingen and in his introduction to his edition of the *Narratio Prima* of Rheticus, which was published together with the *Mysterium Cosmographicum* in 1596 (KGW *1*, p. 84, l. 47).

As we have seen, Kepler describes his theory as originating from his perception of a possible cosmological significance in a well known mathematical fact that had been presented to him in an astronomical or astrological context. He makes rather little of the transition from the unsuccessful use of polygons to the idea of using polyhedra, and the transition does, indeed, seem a very natural one, in view of the close mathematical analogy between the circumcircle and incircle of a polygon and the circumsphere and insphere of a polyhedron. Moreover, the fact that it was usual to imagine planetary orbs as spheres must have played a part in reminding Kepler that, although he was inclined to think of the orbits as merely closed curves, the problem was in fact a three-dimensional one. (It is not clear whether at this stage Kepler had guessed, or was willing to assume, that the orbits were actually plane curves.) Once he had thought of using polyhedra, the five Platonic solids were an entirely obvious choice. Not only were there exactly five of them, but also Kepler, like many of his contemporaries, followed Proclus in believing that the *Elements* had been written with the aim of establishing the existence and properties of these five polyhedra, which were in some sense 'World Figures', either describing or determining some fundamental property of the corporeal world (see, for example, *Harmonices Mundi* Book II, sect. XXV, KGW *6*, pp. 80–2).

The simple idea from which Kepler claims to have started is so simple that it would hardly need to be ascribed to a source. We surely must accept Kepler's account, unless we are to reject the whole of it and ask instead where Kepler could have come upon the more complex idea of turning to the Platonic solids for an explanation of the structure of the Copernican planetary system. The general idea seems to be derived from Proclus, with references back to Plato and the Pythagoreans. Indeed, in the passage from the *Harmonice Mundi* referred to in the last paragraph[13] Kepler seems to wish to ascribe even the details of his theory to the Pythagoreans (whom he took for Copernicans, of course) but he does not give any reference to written evidence for their having constructed such a theory. Kepler often does give references, and his silence in this case must, surely, be construed as significant – as signifying that, as he said in his note of 1621, he knew of no earlier account of the theory to which he could refer. Whether Kepler had somewhere read something which somehow suggested the theory to him without his being aware of the source is a speculation the present writer is content to leave to others. It seems rather perverse to doubt Kepler's apparently straightforward account of how he came to his theory.

Kepler's arguments in favour of his theory

As we have seen, Kepler presents his theory as a confirmation of the Copernican description of the Universe. This naturally affects the structure of the *Mysterium Cosmographicum* to the extent that the work begins with a description of the Copernican system, but by far the greater part of the book is in fact taken up with Kepler's justification of his own theory.

He gives three types of argument in favour of the theory: arguments from the mathematical properties of the Platonic solids, arguments from the connection of the Platonic solids with astrological phenomena, and arguments from the agreement between the spacing of the orbs calculated from the theory and the spacing deduced from the parameters of the orbits given in *De Revolutionibus* and in the *Prutenic Tables* (Tübingen, 1551). These arguments are presented in the order in which they have just been listed. This explains why Kepler at first gives only a sketchy account of his theory, in

Chapter II, postponing the more detailed description until Chapter XIV, which marks the beginning of his discussion of the agreement with observations. Chapters III to VIII (KGW *1*, pp. 29–34) are taken up with mathematical arguments; Chapters IX to XII (KGW *1*, pp. 34–43) with astrological ones; Chapter XIII describes how to find the radii of the circumspheres and inspheres of the five polyhedra; the rest of the book, Chapters XIV to XXIII (KGW *1*, pp. 47–80), deals with astronomical matters, and ranges rather wider than mere comparison between the values calculated from the theory and the values deduced from observations. Much of this astronomical section is, in fact, concerned with problems to which Kepler was to return, armed with Tycho's observations and an apparently limitless supply of willpower. This part of the *Mysterium Cosmographicum* is analysed fairly briefly but very lucidly by Dreyer (1953). Kepler's own comments, in the 1621 edition of the *Mysterium Cosmographicum*, may be exemplified by his note on Chapter XV (a chapter which Dreyer characterised as 'masterly'): 'The problem caused by this as it were dislocation of the Planetary system [i.e. the fact that the centres of the eccentrics were scattered round the Sun, and movable] and how it can be solved by Brahe's observations of Mars are matters I have described in detail in my Commentaries on the motion of the planet . . .' (KGW *8*, p. 90). (The full title of the book now usually known as the *Astronomia Nova* (Heidelberg, 1609) begins *Astronomia Nova* αἰτιολογητος *seu Physica Coelestis, tradita commentariis de motibus stellae Martis,* . . . and Kepler always calls the work something like 'my commentaries on Mars'. Possibly he felt that the first two words seemed to claim too much if they were taken in isolation from the rest.)

The order of Kepler's arguments

The order in which Kepler presents his arguments, starting with mathematical ones and only later coming to the agreement with observations, reveals his concern to give 'physical or, if you prefer, metaphysical reasons' in favour of his theory, and hence in favour of the Copernican system. The order of the arguments should not be seen as strange, even by modern standards. Modern scientists are, of course, trained to ask at once how well the theory accounts for the observations, but that is because the logically prior question of whether the theory is in a form acceptable for a scientific theory is usually taken as already answered in the affirmative. In the sixteenth century it was natural to answer this first question first. As we shall see, the weakness of Kepler's theory in fact lies in its philosophical basis and not, as might perhaps have been expected, in its power to account for the observations. Kepler continued to believe in his theory because he was confident that an incorrect theory must eventually be shown to be incorrect on observational grounds. This belief is one of his 'modern' traits.[14] Crushing observational evidence against the theory described in the *Mysterium Cosmographicum* appeared only with the discovery of the planet Uranus in 1781, although Kepler had briefly supposed that Galileo had provided such evidence, when early reports of the *Sidereus Nuncius* made it seem possible that Galileo had discovered new planets rather than, as Kepler was to call them, 'satellites' of Jupiter (see Chapter IV below).

The little phrase 'physical or, if you prefer, metaphysical reasons' seems to require comment. Kepler habitually refers to his appeals to Archetypes, Ideas coeternal with the Creator, as being 'physical' reasoning. He appears to be using the word in something close to its etymological meaning – φυσικος derived from φυσις – that is, to mean 'pertaining to nature', in the sense that a 'physical' reason is describing the way that things work in the natural world (the natural world being taken to include celestial as well as terrestrial phenomena). Maestlin does not appear to have suggested that his pupil should explain the point more explicitly, and there is no note on the word in the 1621 edition. The usage does, however, appear to be personal to Kepler (see Jardine, 1979).

Mathematics

Before presenting mathematical arguments for his theory, Kepler first explains why he felt a mathematical discussion was necessary. The explanation is summarised in the first sentence of Chapter III: 'I think it might appear fortuitous, and not the consequence of any cause, that the distances between the six Copernican orbs are such that these five bodies will fit into them, were it not that there is an order among the bodies themselves, and it is in this order that I have placed them between the orbs' (KGW *1*, p. 29). After a short elaboration of this summary, he then proceeds to deduce the order among the Platonic solids.

The first thing he establishes, in Chapter III, is that the solids are to be divided into two kinds, primary and secondary. He lists seven features which distinguish the primary from the secondary solids.

1. Each of the primary solids has a different type of face, whereas the secondary solids all have the same kind, a triangle.
2. Each primary solid has a type of face peculiar to itself, the secondary solids have the triangular face of the tetrahedron.
3. The primary solids have simple solid angles at which three faces meet, in the secondary solids more than three faces meet at each vertex. (Euclid defines a solid angle as being made up of at least three plane angles, in Book XI of the *Elements*.)
4. The primary solids do not owe their origin and properties to anything else, the secondary solids can be seen as derived from the primary ones. (The method of derivation is shown in the diagrams which accompany Kepler's discussion of the same relationship in *Harmonices Mundi* Book V Ch. I and in Book IV of the *Epitome* (Linz, 1619 and 1620, respectively). It consists of making the centres of the faces of the 'primary' solids vertices of the 'secondary' ones (see figure 3.11). It is clear from the discussion in the *Epitome* that Kepler realised that the 'primary' solids could be inscribed in the 'secondary' ones in the same way as the latter are shown to be inscribed in the former in his diagrams. That is, to put it in modern terms, he recognised that the relationship of duality was reciprocal. However, even in the *Epitome* he still uses the relations shown in the diagrams as a way of distinguishing between 'primary' and 'secondary' solids.

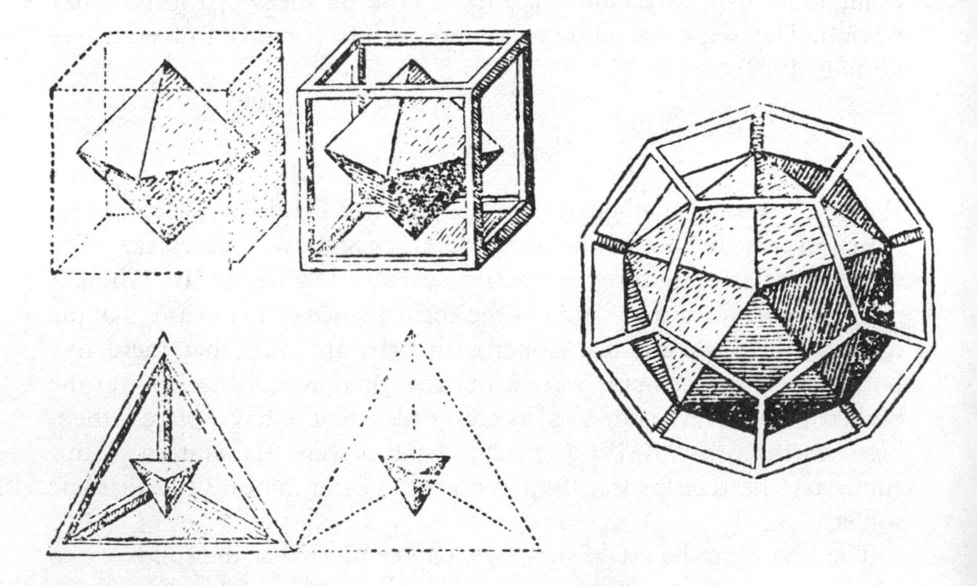

Figure 3.11 Primary and secondary solids (HM Bk V Ch. I)

This departure from mathematical rigour is of a piece with his unwillingness to accord his new regular polyhedra the same status as the Platonic solids – for which see Chapter V below and Field (1979a).)

5. The primary solids do not move harmoniously (*concinne*) unless a diameter is drawn through the centre of one face or the centres of a pair of opposite faces, the secondary solids move harmoniously about a diameter drawn through two vertices. (This probably refers to what would now be called rotational symmetry. The 'primary' solids have only 3-fold rotational symmetry about an axis of symmetry through a vertex, whereas about an axis through the centre of a face they have n-fold rotational symmetry, where n is the number of sides of the face of the solid concerned. The 'secondary' solids have 3-fold rotational symmetry about an axis through the centre of a face, and n-fold rotational symmetry about an axis through a vertex, where n is the number of faces that meet at each vertex. What Kepler calls the 'harmony' of the motion would consist in its degree of symmetry corresponding to the number of sides of each face – the distinguishing feature of each 'primary' solid – and to the number of faces which meet at each vertex – the distinguishing feature of each 'secondary' solid. This is an analogue for polyhedra of the idea that a sphere expresses its own shape by rotation – see Copernicus *De Revolutionibus*, Book I, Ch. IV, NKG, vol. II, p. 12, l. 28.)

6. It is appropriate for the primary solids to stand on one face and for the secondary ones to be suspended by a vertex, that is, the shapes of the solids are best appreciated if they are seen in these positions. (This seems to be no more than a restatement of property 5 in terms of bodies at rest.)

7. There are three primary solids and two secondary ones, three being a perfect number, two an imperfect one. Also, the primary solids between them show all three kinds of angle – the right angle in the cube, the acute angle in the tetrahedron and the obtuse angle in the dodecahedron – whereas the secondary solids share the obtuse angle, though the angles in the octahedron are of all three types – the dihedral angle being obtuse, the angle between the two opposed edges meeting at a vertex being a right angle, and the solid angle at the vertex being acute. (*Mysterium Cosmographicum*, Ch. III, KGW *1*, p. 29.)

Modern mathematical terms have been employed in the above paraphrase since it seems that Kepler had precise mathematical properties in mind, though his expression of them was rather informal ('hand waving'). However, the lack of mathematical formality in Kepler's exposition serves to emphasise his indebtedness to *Timaeus* rather than the *Elements* for this kind of mathematics, whose concern with symmetry is as much aesthetic as purely mathematical, as too is its emphasis on different degrees of perfection. This admixture of aesthetic principles is, in fact, characteristic of much of Kepler's mathematical reasoning, reflecting his belief that God made the Universe as beautiful as possible (see, for example, *Myst. Cosm.*, Ch. IV, KGW *1*, p. 30, l.9). In the present case, Kepler is concerned with the Platonic solids precisely because they are the simplest, and thus the most beautiful, of the polyhedra, those whose properties most closely approach the properties of the sphere. He makes this point explicitly in a long note on Chapter II in the 1596 edition of the *Mysterium Cosmographicum*, a note which was not modified in the second edition (KGW *1*, p. 48). The dominance of aesthetic considerations is therefore exactly what we should expect at this point in Kepler's reasoning.

The chapter ends with the conclusion that 'nothing could be more suitable' than that orb of the Earth should serve to divide the two regions whose structures are determined by the 'primary' and the 'secondary' solids, and Kepler then goes on to consider why there should be three solids outside the orb of the Earth and two inside it. The short chapter in which this question is answered defies adequate summary. Much of it is concerned with God and His intentions in creating the Universe 'For I think that very many causes of what we find in the world can be deduced from the love God bears to Man' (*Mysterium Cosmographicum*, Ch. IV, KGW *1*, p. 30, l. 8). One expression of this love is to place the Earth in the middle of the planets – the Sun for this purpose being counted as a planet, while the Moon is not. (It should perhaps be remembered at this point that, because of the very low value for the distance between the Earth and the Sun that was accepted as correct at this time, the Earth was believed to be relatively larger than we now know it to be.)[15] We thus require three solids to surround the orb of the Earth and two to lie inside it. This accords with Kepler's belief that 'to contain is more perfect, as being active, and to be contained, as being passive, is more imperfect; and the primary solids are indeed more perfect than the others . . .' (KGW *1*, p. 30, l. 31). He has thus explained why there are three planets moving outside the orb of the Earth and two inside it.

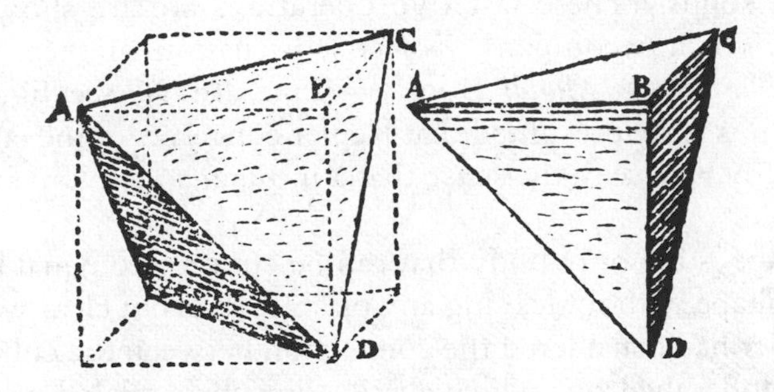

Figure 3.12 Cube to tetrahedron, (*Harm. Mundi* Bk V, Ch. I)

there are three planets moving outside the orb of the Earth and two inside it.

Kepler goes on to consider the order in which the Platonic solids should be placed between the planetary orbs. Since 'to contain is more perfect' this means that we must deduce the order of perfection of the solids and then use them in this order, working inwards from the orb of Saturn. Kepler regards perfection as a matter of mathematical simplicity and his treatment of the cube, the most perfect solid, will serve to exemplify the character of his arguments. We are presented with nine distinguishing features.

1. The cube is the only body to be generated by its face, for the other four are not so generated but are either cut out from the cube – as is

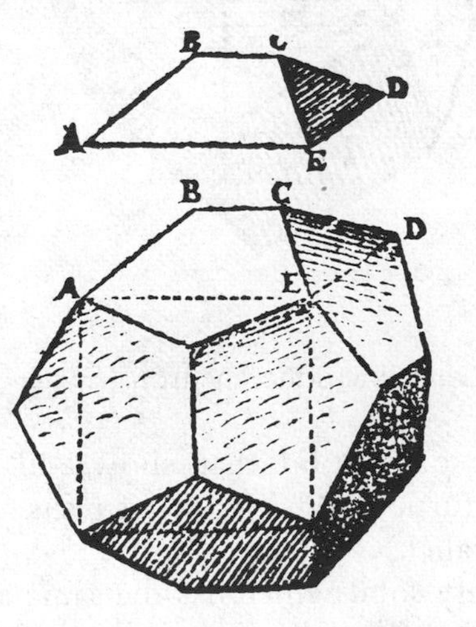

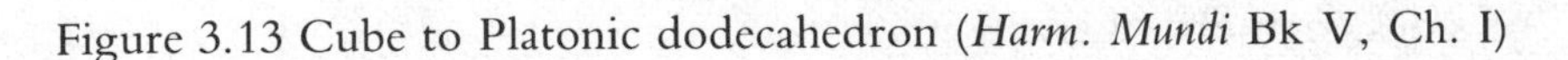

Figure 3.13 Cube to Platonic dodecahedron (*Harm. Mundi* Bk V, Ch. I)

five-faced solids. (These last two operations are are shown in the diagrams which accompany Kepler's discussion of these relationships in *Harmonices Mundi* Book V (Linz, 1619), see figures 3.12 and 3.13. It is to be presumed that Kepler conceives of the cube being generated by its face in the sense that a moving square can sweep out a cube.)

2. The cube is the only body that can be cut up into equal bodies of the same shape without leaving any prisms. (It is not clear whether in 1596 Kepler had considered the connection between the cube and the rhombic dodecahedron, which shares the cube's related property of filling space and is described and illustrated in the *Epitome*, Book IV (KGW 7, p. 270), see figure 3.14. The rhombic solid is much less perfect than the cube since its faces are not even regular.)

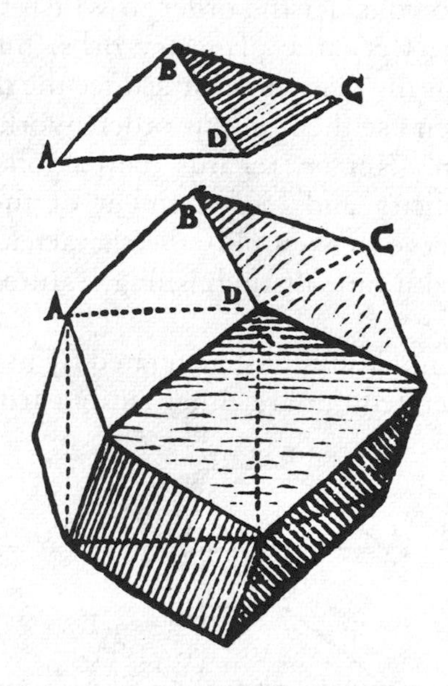

Figure 3.14 Cube to rhombic dodecahedron (*Epitome* Book IV, p. 461)

3. The cube is the only solid that shows all three dimensions whichever way it is turned. (That is, all its parts are either mutually perpendicular or parallel.)

4. Hence it is the only solid which has the same number of faces as the third dimension (*sic*) has limits, that is six, and twice that number of sides, namely twelve.

5. It has the same angle everywhere, that is, a right angle. (This seems to refer to the fact that the dihedral angles and the plane angles of the face are all right angles.)
6. Thus it is the only body which exemplifies the criterion of perfection in the number three which Simplicius in his commentary on Aristotle's *On the Heavens* Book I, Chapter I, quotes from Ptolemy, namely that no more than three straight lines can be mutually perpendicular at any point.
7. It is the simplest of all polyhedra, and is used as the measure of volume. (Kepler adds that it is a natural fact rather than a matter of human convention that any solid body can be imagined as made up of small cubes.)
8. Inscribing a right angle in a circle is a necessary step to obtain many other results, such as inscribing a triangle or a pentagon in a circle. (The right angle, as we have seen, is characteristic of the cube and its significance therefore presumably reflects back onto the cube.)
9. Man himself, the most perfect of the animals, has something of the form of a cube, in that he has, as it were, six surfaces: upper, lower, back, front, right and left. (This is reminiscent of squared-off figures one finds in books that explain how to draw complicated objects in perspective, for example in Dürer's *Underweysung der Messung mit dem Zirkel und Richtscheyt*, Nuremberg, 1525.) (*Myst. Cosm.*, Ch. V, KGW *1*, p. 31).

These nine features which serve to establish that the cube is the simplest and most basic of the Platonic solids are, for the most part, rather more narrowly mathematical than the criteria used to divide the solids into 'primary' and 'secondary' bodies. However, they show the same tendency to specialise rather than generalise: numbers 3, 4 and 6, and perhaps 9 as well, could quite easily have been gathered together as one item, as could numbers 5 and 6 in the earlier sequence. In both sequences, Kepler's attitude is that of the applied rather than the pure mathematician. He is looking for specific instances of mathematical properties that seem to him to have a significance beyond the mathematical entities involved. He is not interested in mathematical truths as such.

The treatment of the tetrahedron, the next most perfect solid, is similar to that of the cube, but the case of the dodecahedron is left as an exercise to the reader: 'what should be thought of its properties will easily become apparent if you compare them with those mentioned for the other solids' (*Myst. Cosm.*, Ch. VI, KGW *1*, p. 33, l.14). It seems as though for Kepler, as for Plato, the dodecahedron merely had to take what was left for it.

The treatment of the two 'secondary' solids is rather shorter than that of the first two 'primary' ones and is to some extent derived from consideration of the positions assigned to the corresponding 'primary' solids. This is, of course, quite reasonable, since the 'secondary' solids are regarded as derived from the 'primary' ones.

Kepler's mathematical justification of his theory thus strongly resembles Timaeus' justification of the theory that related the polyhedra to the elements. Though it is much longer than its prototype it consists only of an array of mathematical facts. The array includes some very simple facts (such as that the 'secondary' solids all have triangular faces), some that are theorems (the construction of a tetrahedron by removing pyramids from a cube is based on Proposition 15 of Book XIII of the *Elements*), and a few that have considerable significance (such as the possibility of dividing any solid into small cubes). Most of the mathematics involved is treated very sketchily and no proofs or even diagrams are supplied. This probably indicates that Kepler expected that many of the facts he mentioned were already well known to his readers.

Only two notes are added to the mathematical chapters of the *Mysterium Cosmographicum* in the 1621 edition. The first refers the reader to a fuller treatment of the same material in Book IV of the *Epitome* (Linz, 1620), and 'a little more' in the first chapter of *Harmonices Mundi* Book V (Linz, 1619) and (where it turns out to be very little) in Chapter XIII of the *Mysterium Cosmographicum* itself. The second note merely concerns the angle in a semicircle. In fact, the two later works appear to add almost nothing to what is said in the *Mysterium Cosmographicum*, except that Book IV of the *Epitome* contains some diagrams (because the work was intended to be introductory or because Kepler was now in a better position to influence his printers?) and the passage in *Harmonices Mundi* Book V describes, and illustrates, one of the two regular star polyhedra which Kepler discovered. It seems we must therefore assume that what he had published in 1596 still seemed valid to him in 1621. This means that he must not only have accepted it as mathematically valid, but must also have believed himself justified in the cosmological conclusions he had drawn from his mathematics.

Astrology and Numerology

Chapters III to VIII of the *Mysterium Cosmographicum* were concerned with establishing that Kepler's theory was mathematically coherent – and did not, for example, involve taking the Platonic solids in some arbitrary order. The astrological and numerological chapters are concerned with what are, to Kepler, more or less observational tests of the theory. He regards the astrological facts as well established, and in Chapters IX, XI, and XII he sets out to show how his theory can account for them. In Chapter X he takes it as well established that certain numbers are of cosmological significance and shows how they may be seen as characteristic of the Platonic solids – arising, for example, as the numbers of sides or faces or angles of the bodies. In a note on this chapter in the second edition of the *Mysterium Cosmographicum* he states explicitly that he regards geometry as prior to arithmetic and refers the reader to *Harmonices Mundi* Book I for a fuller account of their relationship. The earlier treatment is very crude indeed compared with the later one and is of interest mainly in relation to it. I shall therefore postpone consideration of this chapter of the *Mysterium Cosmographicum* until the discussion of the relevant part of the *Harmonice Mundi* in Chapter V below.

Kepler's astrological beliefs changed fairly considerably between 1596 and 1621, and this change is reflected in the notes on the astrological chapters in the second edition of the *Mysterium Cosmographicum*. As we have already noted, Chapter XI, which deals with the origin of the Zodiac, is dismissed in the first note as being of no interest, presumably in the sense of being of no interest in connection with the matter in hand (that is, the relation of the Platonic solids to the structure of the Universe) since Kepler did see fit to add another twelve notes after this first one. Chapter XII, which deals with astrological Aspects, is given thirty-nine notes, and the total length of the notes is rather greater than the original length of the chapter. The material of this chapter is connected with music, that is, musical ratios, and with the division of a circle into equal parts by inscribing a regular polygon in it. These problems are the subjects of two of the five books of the *Harmonice Mundi* and discussion of Kepler's earlier treatment of them will thus be postponed until we consider the later one (see Chapter V below).[16]

The first of the astrological chapters of the *Mysterium Cosmographicum*, Chapter IX, explains the powers of the planets with reference to the corresponding polyhedra. This chapter is only very lightly annotated in the second edition of the work. There is a general note, which says that this chapter should not be taken as part of the main work but as an astrological digression. Kepler nevertheless invites the reader to compare his reasons with those given by Ptolemy in the *Tetrabiblos* and the *Harmonica* (KGW *8*, p. 59). The two notes which follow are both short, and deal with points of only minor importance as far as the main subject of the chapter is concerned. It would therefore seem that although Kepler had modified some of his astrological ideas since writing the *Mysterium Cosmographicum* he still believed it was reasonable to explain the 'observed' powers of the planets in terms of the mathematical entities that had been invoked to explain the spacing of their orbits. Each planet is associated with the polyhedron which was used to account for the spacing between the sphere of the planet concerned and the sphere of the planet which lies next to it in the direction towards the Earth. This leads to the correspondences

Saturn	⟷	Cube
Jupiter	⟷	Tetrahedron
Mars	⟷	Dodecahedron
Venus	⟷	Icosahedron
Mercury	⟷	Octahedron.

Kepler merely states that this seems very reasonable, without presenting any arguments. Since astrology deals with the effects of other planets upon the Earth, even Kepler's astrology was always to some extent geocentric. This presumably accounts for his relating the planets to the Earth in the way he does here. The mathematical properties of the polyhedra which were described in earlier chapters are now used to explain astrological properties of the planets. For example, the fact that the solids corresponding to Jupiter, Venus and Mercury all have the same face, the triangle, is the cause of their friendship for one another (KGW *1*, p. 35, l.22), and the fact that the square, which is characteristic of Saturn, being the face of the cube, is also to be found inside the octahedron, the body associated with Mercury, 'makes peace between their habits' (KGW *1*, p. 35, l.28). The chapter consists almost entirely of a series of very brief assertions like these. Since the astrological powers of the planets are for the most part rather less ponderable than the physical properties of the elements, often being expressed in terms of human character traits and human relationships, Kepler's mathematical explanations do not on the whole seem as appropriate to his subject as Timaeus' explanations did to his. Both sets of explanations are, however, of the same kind. The mathematical forms, by means of which Timaeus explains the properties of the elements, have no existence in the physical world, and nor do the polyhedra which Kepler's diagrams show as lying between the orbs of the planets.

Since Kepler's note of 1621 refers to this chapter as a digression, we must presume that although he did not, apparently, wish to modify it as drastically as he had modified the other chapters in this section, he nevertheless no longer regarded it as an important part of the justification of his theory. It is very possible that he had never regarded it in that light, as is, indeed, suggested by the brevity of his treatment. We should not, however, take this brevity to suggest that Kepler was not very interested in astrology as such. There is ample evidence to the contrary. Kepler wrote a treatise on the subject, *De Fundamentis Astrologiae Certioribus* (Prague, 1602), and the work, being written in Latin, is clearly intended for the learned. Much of *De Stella Nova* (Prague, 1606) is taken up with astrological matters, such as the significance of the fiery trigon. Book IV of the *Harmonice Mundi* (Linz, 1619) is entirely concerned with astrology. Moreover, Kepler repeatedly discusses astrological problems in his correspondence.[17] His summary treatment of astrological arguments in the *Mysterium Cosmographicum* is probably to be explained by the fact that although he regarded astrology as being founded on a solid basis of observations (see, for example, Chapter IX of *De Stella Nova*, KGW *1*, p. 189 ff) he nevertheless did not regard it as an exact science on a level with astronomy. In this he was in agreement with the opinion Ptolemy expresses at the beginning of his introduction to the *Tetrabiblos*. As Simon has shown, Kepler tried to reform the accepted astrology of his day in much the same way as he tried to reform its astronomy, but in the former case, as Simon puts it 'The remedy was clearly one of those which in the end kill the patient'.[18]

Astronomy

The astronomical chapters of the *Mysterium Cosmographicum* give a very full treatment of the problem of testing Kepler's theory against values deduced from observations. It is crucial to this problem that it is not possible merely to compare theory with observation. In Kepler's day, as in our own, the physical facts of astronomy tend to be deducible from observational facts only with the aid of much theory, and even the simplest of the facts which Kepler's theory sets out to explain are in their turn theory-laden. It is only in a Copernican Universe that there are six planets, explained by the five polyhedra, and only in a Copernican Universe will the orbit of the Earth be seen as dividing the orbits into a set of three and a set of two, explained by the division of the polyhedra into 'primary' and 'secondary' solids.

When Kepler comes to check how well the five polyhedra fit between the planetary spheres, the intrusion of astronomical theory is much more subtle, but equally inexorable. As Grafton (1973) has shown, when Kepler was engaged in writing these chapters of the *Mysterium Cosmographicum* he repeatedly found that his grasp of the relevant parts of Copernicus' work was not adequate to the task. However, Maestlin was apparently very willing to continue to act as Kepler's teacher, and Kepler's technical questions about Copernican planetary theory received long technical replies, one of which was eventually adapted to form an appendix to the printed version of the *Mysterium Cosmographicum*. Unlike Copernicus, Maestlin does not deal with each planet separately. He begins with 'The theory of the Sun, or rather of the great sphere of the Earth', and then turns to the theory of the Moon and then to the 'superior' planets, now the 'outer' planets, Saturn, Jupiter and Mars, after which he deals separately with Venus and Mercury. Most of his numerical values are taken from the *Prutenic Tables* but he also uses values from *De Revolutionibus*, giving references to particular chapters of the latter work. Perhaps Maestlin's influence lay behind Kepler's convenient habit of regularly giving references in a usable form.

For the purposes of the very simple comparison between Kepler's theory and Copernicus' dimensions of the planetary orbs which is described in Chapter XIV of the *Mysterium Cosmographicum*, we may ignore almost all of the Copernican apparatus that is described by Maestlin in his Appendix (and in rather more explicit detail by Neugebauer, 1968). What we need to know is summarised in the diagram, to which we have already referred, showing the orbs round the Sun (Tabella IIII, see figure 3.6). As noted above, the key to this diagram, whose wording may well be Maestlin's, states that the common centre of all the spherical shells is G, the centre of the Great Orb, 'the centre of all [the spheres] and near to the actual body of the Sun', though, as Aiton (1977, 1981) notes, the shells defining the orb of the Earth have in fact been centred on the Sun. The planetary orbs are explicitly defined with reference to the actual paths of the planets in space: '. . . it seems to us that we do not need any orb that extends beyond the path of the planet (*ultra viam planetae*) in order to account for its motion' (KGW *1*, p. 47, l. 36). Kepler apparently sees this as a conclusion to be drawn directly from the Copernican description of the planetary system (KGW *1*, p. 47, l. 34). For the purposes of the *Mysterium Cosmographicum* it led him only to an unorthodox definition of a planetary sphere, but its emphasis on the actual path of the planet points forward to the derivation of the laws of planetary motion in the *Astronomia Nova* (Heidelberg, 1609).

Kepler compares the dimensions of these planetary spheres, calculated from the numbers given by Copernicus, with the dimensions that would be obtained if the appropriate Platonic solid were taken to determine the outer surface of the next sphere from the 'known' inner surface of the one outside it. That is, he checks each of the five 'known' values of the ratios of the radii of the relevant spheres against the value calculated from the appropriate Platonic solid. We thus have five independent comparisons. Kepler puts them in a form which is not quite a table (KGW *1*, p. 48). It reads

If the inner radius for the sphere of		is 1000. The outer radius should be for the sphere of			And is, according to Copernicus,	Book V of Copernicus	
	♄		Jupiter	577		635	ch. 9
	♃		Mars	333		333	ch. 14
	♂		Earth	795		757	ch. 19
	Earth		Venus	795		794	ch. 21 & 22
	♀		Mercury	577 or 707		723	ch. 27

The first column of figures is calculated from the Platonic solids. Kepler carried out the required calculations in Chapter XIII, and has rounded his numbers off to three significant figures (see table 3.2). The first five numbers are the ratios of the radii of the circumspheres to the radii of the inspheres. The second number for Mercury is obtained by taking the inner sphere for the octahedron as the sphere which touches the edges of the solid. (Kepler calls this number the radius of the circle touching the square of the octahedron, and he gives no indication of knowing that spheres can be constructed to touch the edges of any other of the Platonic solids. Such a sphere is called the 'midsphere' of the solid concerned.)

Table 3.2 *Radii of circumspheres and inspheres of the Platonic solids*

polyhedron	R_c/R_i	R_i if $R_c = 1000$ (to 3 sig. fig.)
cube	$\sqrt{3}$	577
tetrahedron	3	333
dodecahedron	$\sqrt{15 - 6\sqrt{5}}$	795
icosahedron	$\sqrt{15 - 6\sqrt{5}}$	795
octahedron	$\sqrt{3}$	577
R_c/R_m	$\sqrt{2}$	707 (R_m)

For the purpose of later comparisons it will be convenient to recast Kepler's results in the form of a table and to add a column of absolute errors and one of anachronistic percentage errors, calculated with an anachronistic electronic calculator and given to the nearest 1% (see table 3.3).

However, this comparison has ignored the Moon. If the thickness of the orb of the Earth is increased to accommodate the orbit of the Moon, the values shown in table 3.3 must be modified as shown in table 3.4. It will be noted that the modification gives slightly closer agreement between theoretical and 'observed' values.

Table 3.3 *Orbs and polyhedra* (*Myst. Cosm.* Ch. XIV)

planet	outer radius of lower orb if inner radius of upper one is 1000		diff. th − obs	
	from polyhedron 'th'	from Cop. 'obs'	abs	% of obs
♃	577	635	− 58	−9
♂	333	333	0	0
⊕	795	757	+38	+5
♀	795	794	+1	0
☿	577	723	−146	−20
	707 (mid)		−16	−2

Table 3.4 *Orbs and polyhedra, including Moon with Earth* (*Myst. Cosm.* Ch. XIV)

planet	outer radius of lower orb if inner radius of upper one is 1000		diff. th – obs	
	from polyhedron 'th'	from Cop. 'obs'	abs	% of obs
♃	577	635	−58	−9
♂	333	333	0	0
⊕ + ☾	795	801	−6	−1
♀	795	847	−52	−6
☿	577	723	−146	−20
	707 (mid)		−16	−2

Kepler's comment on the agreement between his columns of figures is

> 'See, the corresponding numbers are close to one another, and those for Mars and for Venus are the same. Those for the Earth and for Mercury are not very different; only those for Jupiter are widely separated, but no-one will be surprised at that for such a large distance. And for Mars and Venus, which lie next to the Earth, you see how much difference is made by the increase in thickness of the orb of the Earth caused by adding the little orb of the Moon though the little orb is hardly one twentieth of the size of the orb of the Earth.' (KGW *1*, p. 48, l. 35 ff).

He concludes that his theory seems to give a good description of the distances.

In his following chapter, Chapter XV, Kepler transfers the centre of the system to the Sun, and once more compares the dimensions of the orbs found from his theory with those derived from observation. The comparison is displayed in the form of two tables, one in which the Moon is ignored and one in which its orb is included in that of the Earth (as in the tables of Chapter XIV). In table 3.5 we have combined Kepler's two tables, adopted decimal fractions in place of his sexagesimal ones, and added columns of percentage errors. Kepler took the mean radius of the orbit of the Earth to be unity, so the distances are, in effect, in modern Astronomical Units (AU).

It will be noted from table 3.5 that the agreement is not markedly different from that Kepler obtained in his previous chapter (see tables 3.3 and 3.4). However, as Aiton (1981) has pointed out, Kepler's tables contain numerical errors – which, moreover, he did not correct in the second edition of the *Mysterium Cosmographicum*, presumably because by 1621 he considered this earlier form of his theory to be of merely historical interest. If we adopt Aiton's corrected values, we obtain the results shown in table 3.6, which allow us to prefer the version of Kepler's theory in which the orb of the Moon is included in the orb of the Earth.

The recalculation of the dimensions of the orbs as measured from the Sun has not greatly affected the degree of agreement between theoretical and observed values. It is therefore not very surprising that Kepler makes

Table 3.5 *Orbs and polyhedra* (*Myst Cosm*. Ch. XV)

		observed	theoretical		errors (th − obs)/obs%	
		distance	no Moon	with Moon	no Moon	with Moon
♄	aph	9.987	10.599	11.304		
	peri	8.342	8.852	9.441	+6	+13
♃	aph	5.492	5.111	5.451		
	peri	4.999	4.652	4.951	−7	−1
♂	aph	1.649	1.551	1.658		
	peri	1.393	1.311	1.398	−6	0
⊕	aph	1.042	1.042	1.102	0,	by definition
	peri	0.958	0.958	0.898		
♀	aph	0.741	0.761	0.714	+3	−4
	peri	0.696	0.715	0.671		
☿	aph	0.489	0.506	0.474	+4	−3
	peri	0.233	0.233	0.219		

Table 3.6 *Orbs and Polyhedra (Myst. Cosm*. Ch. XV) with corrections from Aiton (1981)

		observed	theoretical		errors (th − obs)/obs%	
		distance	no Moon	with Moon	no Moon	with Moon
♄	aph	9.727	10.011	10.588	+3	+9
	peri	8.602	8.854	9.364		
♃	aph	5.492	5.109	5.403	−7	−2
	peri	4.999	4.650	4.918		
♂	aph	1.648	1.550	1.639	+6	−1
	peri	1.393	1.310	1.386		
⊕	aph	1.042	1.042	1.102	0,	by definition
	peri	0.958	0.958	0.898		
♀	aph	0.721	0.762	0.714	+6	−1
	peri	0.717	0.757	0.710		
☿	aph	0.481	0.535	0.502	+11	+4
	peri	0.233	0.260	0.242		

no comment upon it, merely turning to the next possible source of modification to the dimensions, namely matters relating to the motion of the Moon (Ch. XVI). After these, he considers the special problem with Mercury (Ch. XVII) and general problems in calculating the radii of the circles from which the orbits are compounded (Ch. XVIII). Chapter XIX then returns to the 'remaining differences' and Kepler points out, not as a general statement, but piecemeal in dealing with the orbits of individual planets, that every orbit is uncertain to some degree because of the difficulty of making the observations required to calculate the orbits. He notes that this is particularly true of observations of Mercury, which is never seen far from the Sun. His reference is explicitly to observations of oppositions and maximum elongations – the types of observation that were traditionally used to calculate the orbital parameters for Ptolemaic as well as Copernican models.

These three chapters make it clear that Kepler was aware that some hard astronomical work would need to be done before better dimensions could be calculated. However, he makes no suggestion that the results obtained in Chapter XV or the problems analysed in Chapters XVI to XVIII should be seen as weakening the conclusion he reached at the end of Chapter XIV, namely that his theory is in reasonably close agreement with the results of observation.

At the beginning of Chapter XX (KGW *1*, p. 68) Kepler says that his satisfactory explanation of the spacing of the planets is his strongest argument for the Copernican theory, but he adds that he hopes that the *motions* of the planets (that is, their periodic times) can also be related to the dimensions of their Copernican orbs – as we have seen, he had set himself the task of explaining 'the Number, Size and Motions of the Orbs' (*Myst. Cosm.*, Preface, KGW *1*, p. 9, l. 33). His first attempted explanation of the periods is by straightforward proportionality, $P \propto r$, his second and more accurate one is exactly analogous to his explanation of the dimensions of the orbs, in that it considers the difference between each orb and the one inside it. As Gingerich (1975) has pointed out in his analysis of this chapter, if Kepler had been the number-juggler that popular accounts describe, he might have found his third law, $P^2 \propto r^3$, at this point in his work. However, in order to have discovered it, he would have had to accept rather large errors. For although the dimensions Kepler gives for the orbs in the *Mysterium Cosmographicum* are not very different from those he was to calculate later from Tycho's observations (and the periods, of course, had been known very accurately for centuries past) the squares of the periods (in years) do not agree strikingly well with the cubes of the mean radii of the orbs (in units of the mean Earth-Sun distance). As can be seen in table 3.7, the errors range from −2.5% (Jupiter) to +23% (Mercury). The evidence for the third law appears considerably weaker than the evidence for the polyhedral archetype.

In any case, Kepler was in fact not juggling numbers, but working from his understanding of the physics of the system, and his theory consequently gives a markedly inexact account of the moving forces.[19] The next chapter, entitled 'What is to be gathered from the discordance',

Table 3.7 *Kepler's third law, using linear dimensions given in Myst. Cosm. Ch. XV, and periods given in Myst. Cosm. Ch. XX.* The figures in brackets show the results obtained if we adopt the corrections made by Aiton (1981) (compare tables 3.5 and 3.6).

	mean radius (r)	r^3	period (P)	P^2	P^2/r^3	error(%)
♄	9.1645	769.7	29.46	867.7	1.127	+13
♃	5.2455	144.3	11.86	140.7	0.9751	−2.5
♂	1.5210 (1.5205)	3.519 (3.515)	1.881	3.538	1.005 (1.006)	+0.5 (+0.6)
⊕	1 by definition		1 by definition		1	
♀	0.7185 (0.7109)	0.3709 (0.3717)	0.6152	0.3785	1.020 (1.018)	+2.0 (+1.8)
☿	0.3610 (0.3570)	0.04705 (0.04550)	0.2408	0.05800	1.233 (1.275)	+23 (+27)

examines the differences between the ratios of the mean radii of neighbouring spheres obtained from Kepler's theory and the ratios obtained from Copernicus' orbital parameters, and makes a very brief attempt to relate each of these differences to the appropriate Platonic solid. Kepler's estimate of his own success may be guessed from the final lines of the chapter, which express the hope that he will have stimulated others to achieve a reconciliation between his figures.

This attempt to use the system of nested orbs and polyhedra to explain the motions of the orbs is the last reference to the theory in the *Mysterium Cosmographicum*, whose two remaining chapters are concerned with technicalities of Copernican planetary theory (Ch. XXII) and with 'The astronomical beginning and end of the world, and the Platonic year' (Ch. XXIII). The final section of the work is a hymn, whose astronomical content appears to be minimal.

Kepler's belief in his system of orbs and polyhedra

The mathematical basis of Kepler's theory was derived from the *Elements*, whose purpose he believed to have been to establish the existence and properties of the five regular solids which he used to explain the structure of the Universe. Their ancient name of 'World Figures' lent its authority to his application of them. The particular mathematical properties which Kepler used in working out the details of his theory are treated in the cursory manner of *Timaeus* rather than the precise manner of the *Elements*, with repeated appeals to the aesthetic principle and very little mathematical exposition. His mathematical justification for the theory is, however, far from cursory. He is at considerable pains to establish that his theory contains no arbitrary elements – it was, after all, the existence of arbitrary elements that he used as a weapon against Ptolemaic astronomy.

As we have seen, Kepler's theory was designed to solve two related cosmological problems which arose in connection with the Copernican system. The simpler problem was that of the new number of the planets, now six instead of the traditional seven, and the apparently arbitrary position of the Earth among them. Kepler's solution to the first part of this problem was entirely clear-cut: there were six planets because there were five Platonic solids. The second part he solved by dividing the Platonic solids into 'primary' and 'secondary' bodies and placing the orb of the Earth so as to divide the nobler bodies from the baser (like an astronomical salt cellar). He also provided non-mathematical justifications for the position of the Earth, referring to God's love for Mankind, which is expressed by the Earth's being in the centre of the planets in much the same way that the Sun was said to be 'in the centre' of the orbs in geocentric Neoplatonic descriptions of the Universe.

The more difficult problem that Kepler's theory set out to solve was the problem of the spacing of the planetary orbs. Since the Copernican theory allowed the actual proportions of the planetary orbs to be calculated from observations, the gaps between them were also determinate. In explaining these gaps Kepler removed another apparently arbitrary feature from the Copernican description of the Universe, and he therefore saw his theory as providing support for that of Copernicus. As we have seen, Kepler's theory was firmly based on the Platonic belief that mathematical truths should be seen as determining the structure of the observable Universe. We have seen also that the theory was in good agreement with the actual numerical values of the dimensions of the orbs as derived from observation. That is, the differences between theoretical and observational values of the dimensions seemed to Kepler to be sufficiently small to suggest that they were due to errors in the observational values. As can be seen from table 3.5, this conviction was entirely reasonable by the standards applied today. It is not, however, clear what standards Kepler himself was applying in 1596.[20]

The fact that Kepler allowed a second edition of the *Mysterium Cosmographicum* to be printed in 1621, in itself suggests that the huge amount of astronomical work which he had done in the meantime had not led him to reject the theory put forward in this early publication. The next chapter will attempt to explain why it had not done so.

IV
Mysterium Cosmographicum 1621

It appears that the learned world's response to the *Mysterium Cosmographicum* was generally favourable, if we may judge by Kepler's comment in *Harmonices Mundi* Book II that '. . . no-one has attacked it [i.e. the theory described in the *Mysterium Cosmographicum*] in these twenty-two years, but even the pupils of Ramus, that hot-headed scholar, the scourge of Euclid, even they have been drawn to it, and it now excites so much interest that mathematicians are calling for a second edition to be brought out'.[1]

The prospect of a second edition of the *Mysterium Cosmographicum* presented Kepler with a problem. As he explained in the new dedicatory letter which accompanied the second edition 'almost every one of the astronomical works I have written since that time could be referred to some particular chapter of this little book, and be seen to contain either an illustration or a completion of what it says, . . '.[2] It seems that no historian has presumed to differ from this judgement – though Frisch chose to rephrase it in a rather more poetic form, calling the *Mysterium Cosmographicum* 'the well from which Kepler's later works were drawn' (KOF I, p. 1). The metaphor does not seem felicitous. The *Mysterium Cosmographicum* was not a 'well' but merely the first result of what was to prove a lifelong interest in the same problem, the problem of explaining 'the number, sizes and motions of the orbs'.[3] The difficulty in producing a second edition of Kepler's youthful contribution to the solution of this very large problem arose from the success of his later contributions. The 'dislocated' Solar system of Chapter XV of the *Mysterium Cosmographicum* had been transformed into an orderly pattern of planets whose elliptical orbits all had the Sun in one of their foci (*Astronomia Nova*, Heidelberg, 1609). The unsatisfactory attempt to relate the periods ('motions') of the orbs to their sizes, in Chapters XX and XXI, had been superseded by the discovery that the squares of the periods varied as the cubes of the mean radii of the orbits (*Harmonices Mundi Libri V*, Linz, 1619, Book V, Chapter III). However, Kepler had not changed his mind about the main subject of his work, the explanation of the number and sizes of the orbs by means of the Platonic solids,[4] and he was therefore prepared to accede to the requests of 'friends, and not only booksellers but also those learned in Philosophy' (KGW *8*, p. 10, l. 6), who asked for a second edition to be printed. Some people suggested that he should rewrite and extend parts of the work. However, Kepler's own feeling was 'that the book could not be brought up to date (*perfici*) except by writing into it, almost in their entirety, most of the works I have published in these twenty years' (KGW *8*, p. 10, ll. 11–13). This would appear to be a somewhat extreme position, but it is easy to sympathise with Kepler's unwillingness to set about writing summaries of large parts of later astronomical works, merely for the purpose of substituting them for the corresponding chapters of the *Mysterium Cosmographicum*. In fact, by the time the second edition of that work was published, approximations to such summaries could be found in the *Epitome Astronomiae Copernicanae* (Linz, 1618–21). Instead of rewriting, Kepler decided to reprint the text of the first edition of the *Mysterium Cosmographicum* and to add notes. This allowed him to make it clear why he had changed his mind on certain matters – a fact which Kepler regarded as an advantage (KGW *8*, p. 10, l. 15 ff). The procedure is thus slightly reminiscent of the spirit in which Kepler wrote the *Astronomia Nova*: 'I recount all my attempts, so as to make it that much more clear why I took this particular path'.[5] The second edition of the *Mysterium Cosmographicum* is roughly half as long again as the original edition: in the twentieth-century re-editions the 1596 edition covers 77 pages and the 1621 edition covers 121.[6]

Although some of the notes give quite detailed explanations of Kepler's changes of mind, many of them merely refer the reader to the other works in which the matter in hand is treated in the manner Kepler now regards as more correct. The tone of such notes is rather that of pleasure in his later success than one of explanation or apology for his earlier failures. For example, on the title of the chapter which deals with the 'discordance' between theory and observation in his suggested relation between the periods of revolution and the sizes of the orbs, he comments 'I now think this guesswork (*coniectatio*) is superfluous. Having found the true proportion, in which the discordance is exactly nil (*defectus plane nullus*), what need have I of this false discordance?'[7]

Many of the notes Kepler added are severely critical. An extreme example is the first note on Chapter II, which begins 'Oh what a mistake (*O male factum*)' and continues largely in the form of rhetorical questions. This tirade is directed against what Kepler believes to be a serious theoretical error, namely the dismissal of lines and surfaces as playing no important part in determining the structure of the Universe. 'For why should we exclude lines from the archetype of the world when in the work itself God has shown lines, namely the motions of the planets?' (KGW *8*, p. 50, l. 7). The reader is referred to the *Harmonice Mundi* for a detailed account of what Kepler now believes to be the truth of the matter.

Most of the notes deal with points that are of some significance, but Kepler does also notice such things as an arithmetical error in Chapter X, commenting 'behold a manifest hallucination; eight is not a factor of sixty' (KGW *8*, p. 60, l. 32). All in all, the number and character of the annotations suggest that Kepler reread his work with considerable care and intended that the annotations should be as complete as possible. There is no sign of any inclination to take an indulgent attitude towards his early work. Indeed one would hardly expect such indulgence on the part of a man who had proved capable of writing off hard-won results with the ruthlessness shown in the *Astronomia Nova*.

We have already seen that many of Kepler's annotations consist of little more than a reference to the book in which the reader may find a more satisfactory treatment of the point concerned. Most such notes refer the reader either to the *Astronomia Nova* (Heidelberg, 1609) or to *Harmonices Mundi Libri V* (Linz, 1619), but two notes refer to the *Dissertatio cum Nuncio Sidereo* (Prague, 1610).[8] Since the *Sidereus Nuncius* had very significant implications for cosmological theory, we shall consider Kepler's reaction to this work before turning to his annotations to the *Mysterium Cosmographicum*.

Sidereus Nuncius (Venice, 1610)

The *Sidereus Nuncius* describes, with some autobiographical anecdote but very little experimental detail, how Galileo constructed a series of telescopes and used the best of them to observe the Moon, the stars, the Milky Way and the planets. He observed that the Moon showed ranges of mountains, like those on the Earth (but, he believed, much higher); that the telescope allowed him to see more stars than were visible to the naked eye (and that the stars did not increase in apparent size as much as he had expected from what he knew of the magnifying power of his instrument); that the Milky Way was, indeed, as some ancient authors had stated, made up of faint stars; and that the planet Jupiter was attended by four small bodies which revolved about it and were carried round with it in its orbit, thus bearing the same relation to Jupiter as Copernicus believed the Moon to bear to the Earth.

Kepler first used a telescope in September 1611, and unlike Galileo he provides a description of the quality of the images obtained with the telescope (which Galileo had presented to the Elector of Cologne):

> '. . . For to us Jupiter too, like Mars, and in the morning Mercury and Sirius, appeared four-cornered (*quadranguli*). One of the diameters through the angles was blue (*caeruleus*), the other reddish (*puniceus*), in the middle the body was yellow (*flavus*), amazingly bright . . .'[9]

This is a thoroughly convincing description of images affected by chromatic aberration, spherical aberration and astigmatism. The performance of the instrument would have been improved by stopping down the aperture. Kepler ascribes the defects in the images to the weakness of the sense of sight when confronted with the great quantity of light accumulated by the instrument. Since he does not mention any device for decreasing the aperture of the telescope (the simplest way of cutting down the quantity of light) it seems probable that Galileo had not supplied the Elector with any stops. Galileo did, however, use stops with his own telescopes, so it is possible that they gave somewhat better images than those so graphically described by Kepler.

The *Sidereus Nuncius* begins with a description of the telescope but it does not give any details as to how Galileo's observations were made. However, the first astronomical section of the work contains a very detailed account of the slowly-changing patterns which Galileo saw on the face of the Moon. The appearances are described very fully before Galileo proffers his explanation that what he has seen are mountains and their shadows. There has been much dispute about Galileo's skill as an observer, and it seems most unlikely that the matter can ever be resolved one way or the other. (Personally, I am inclined to accept Galileo's description of his observations of the Moon, and the rest of his book, as an essentially accurate record of what must have been a very exciting series of observations.[10]) In any case, our concern here is not with historians' opinions of the *Sidereus Nuncius*, but with Kepler's opinion of it. It is clear from the fact that he wrote the *Dissertatio*, and from an explicit statement near the beginning of the work, that Kepler took the *Sidereus Nuncius* to be a true record of observations made by Galileo.

Dissertatio cum Nuncio Sidereo (Prague, 1610)

The events surrounding the composition of Kepler's letter to Galileo and its later publication, in a slightly modified form, as the *Dissertatio cum Nuncio Sidereo* (Prague, 1610, KGW *4*, pp. 283–311), have been examined in some detail by Rosen in the introduction and notes to his translation of the work, *Kepler's Conversation with Galileo's Sidereal Messenger* (New York and London, 1965).

Rosen has established, from the autobiographical material contained in the work itself and from associated correspondence, that Kepler wrote his letter between 13 April and 19 April. The evidence that it was, indeed, only on 13 April that Kepler learned of Galileo's request for such a letter is convincing, but it does not seem to be impossible that Kepler, who was a prolific letter-writer, had begun to write to Galileo as soon as he had read the *Sidereus Nuncius*, a copy of which came into his hands on 9 April. In any case, the letter was certainly written quickly, as Kepler himself remarks,[11] and the printed version appeared early in May.

It was not, in fact, from the *Sidereus Nuncius* itself that Kepler had obtained his first news of Galileo's discoveries. About 15 March news of them had been brought to Prague by couriers from Venice and Wackher von Wackenfels had passed the news on to Kepler, in some excitement.[12] Wackher's excitement was occasioned by his belief that he had acquired a powerful new argument to use in his long-standing dispute with Kepler as to whether the Universe might be infinite: the news he had heard, which had originated from people who had seen Galileo's work before it was printed, was that Galileo had discovered four previously unknown planets. Wackher thought these planets must orbit one of the fixed stars, thus indicating that Bruno had been correct in his suggestion that all the stars were Suns (KGW *4*, p. 289, l. 14). Kepler, of course, had no wish to jump to any such conclusion. His first thought was for his system of planetary orbs and regular polyhedra: 'So I asked myself how there could be any increase in the number of the planets without damage to my 'secret of the Universe' which I published thirteen years ago, and in which the five figures of Euclid, which Proclus, following Pythagoras and Plato, calls 'Cosmic figures', allow no more than six planets round the Sun' (KGW *4*, p. 288, l. 34 ff). It then occurred to him that Galileo might have found not planets like the Earth but smaller bodies, like the Moon, one going round each of the planets Saturn, Jupiter, Mars and Venus. He thought that Mercury's moon would be invisible because Mercury was too close to the Sun to be observed clearly (KGW *4*, p. 289, l. 9). Wackher and Kepler agreed to differ, and waited eagerly for Galileo's book to arrive (KGW *4*, p. 289, l. 24).

Kepler recounts this story by way of introduction, and then goes on to give what is more or less a 'commentary' on Galileo's book. The plan of Galileo's book thus dictates the plan of Kepler's, so it is not until the last section of the work that he returns to the subject of the four new planets, which he considered 'the most wonderful thing in your [i.e. Galileo's] book' (KGW *4*, p. 301, l. 1). When he does so, however, he is again at pains to relate them to his 'secret of the Universe'. As he had remarked at the end of the Notice to the Reader, he felt free to defend some of his own ideas (*dogmata*) as well as Galileo's, believing them to be correct (KGW *4*, p. 287, l. 10 ff).

He introduces his account of the 'secret' with a characteristic statement on the status of geometry: 'Geometry is one and eternal, shining in the mind of God. That share in it accorded to men is one of the reasons (*causae*) that Man is the image of God' (KGW *4*, p. 308, ll. 9–10). He then uses the fact that the five Platonic solids determine the structure of 'our planetary world' to prove that if there are other worlds then our world is the most perfect one (KGW *4*, p. 308, ll. 13–33).

Kepler has already noted that since the four little planets round Jupiter have only just been discovered by Man they clearly cannot have been made for his sake (KGW *4*, p. 306, l. 12 ff) but were very probably made for the sake of inhabitants of Jupiter (KGW *4*, p. 307, l. 18). This naturally raised the question of Man's special place in the Universe (KGW *4*, p. 307, l. 32) and Kepler's account of the 'secret' is therefore designed to emphasise the special status of the Earth, which had not been given quite so much emphasis in the full account of the 'secret' in the *Mysterium Cosmographicum*. Apart from this slight change, the theory is apparently exactly as before, though Kepler's account of it is in philosophical terms rather than mathematical ones. For example, the justification of the central position of the Sun is by a series of assertions concerning its nobility and power, assertions which are more reminiscent of *De Revolutionibus*, Book I, Chapter 9, than of the first chapter of the *Mysterium*

Cosmographicum. The nobility of the Earth, less only than that of the Sun, is shown by its being in the middle of the orbs, and by the fact that its orb divides those orbs whose spacings are determined by the primary solids from those whose spacings are determined by the secondary ones. This reasoning is so clearly a non-mathematical summary of the *Mysterium Cosmographicum* that one wonders why Kepler does not give a reference to the work. There are numerous references to others of his works in the *Dissertatio*. Perhaps he felt that a reference to the *Mysterium Cosmographicum* would amount to an untactfully clear reminder to Galileo of what, in the second paragraph of the *Dissertatio*, Kepler had tactfully chosen to call their 'interrupted correspondence', namely Galileo's failure to reply to Kepler's letter asking for his opinion of the *Mysterium Cosmographicum*?[13]

Having thus sketched out his system, Kepler then suggests that the four moons of Jupiter compensate the inhabitants of that planet for their relatively indistinct view of the four planets that lie between them and the Sun (KGW *4*, p. 309, l. 29). However, he also gives a mathematical explanation of the fact that there are four moons of Jupiter, namely that there are three polyhedra which can be constructed from rhombs: the cube (whose face is a square, which is a special case of a rhomb), a twelve-faced solid (related to the cuboctahedron) and a thirty-faced solid (related to the icosidodecahedron) (KGW *4*, p. 309, l. 34). The paraphrase reproduces the brevity of Kepler's treatment. He gives no further explanation, and no reference to any other work, not even one that is in progress. The last two rhombic solids are mentioned again in *De Nive Sexangula* (Prague, 1611), though the reference to the triacontahedron in this work is very much in passing, Kepler's main concern being with the rhombic dodecahedron. The two Archimedean solids to which the rhombic solids have been related and all three of the rhombic solids reappear in *Harmonices Mundi* Book II where Kepler proves that there are exactly three such rhombic solids.[14] A more detailed examination of Kepler's concern with the rhombic solids will be found in Appendix 4.

Since at the time he wrote the *Dissertatio* Kepler did not yet know the dimensions of the orbits of the moons of Jupiter, he could not check whether the rhombic solids would fit between their orbs in the same way as the Platonic solids fitted between the orbs of the planets. This omission was, however, repaired in the *Epitome*,[15] where Kepler not only claimed that the ratios of the radii of the orbs agreed with the ratios of the radii of the spheres associated with the rhombic solids (see Appendix 4), but also showed that the periods of revolution of the moons conformed to his third law. In its final form, therefore, Kepler's application of the rhombic solids to the system of Jupiter's moons strongly resembles his application of the Platonic solids to the Solar system of planets. He does not, however, provide any explicit mathematical justification of the new idea. Perhaps he felt a sympathetic reader could devise his own counterpart to the mathematical chapters of the *Mysterium Cosmographicum*?

Kepler's suggestion that the three rhombic solids might explain the four moons of Jupiter indicates that in 1610, as in 1596, he felt confident that properties of mathematical entities might be seen as determining the properties of the observed physical world. Moreover, he was clearly confident not only of the principle behind the theory described in the *Mysterium Cosmographicum* but also of the correctness of the theory itself, for he suggests that if Galileo finds moons which orbit Mars and Venus then their orbits will not only fit into the scheme of orbs and polyhedra but will even improve the agreement between the observed planetary orbs and those calculated from the theory: 'Recently, while recalculating the orbits and motions of Mars, the Earth and Venus from Brahe's observations, I noticed the spaces between the orbs are slightly too large, so that when the vertices of the dodecahedron are placed as far out as the perihelion of Mars the centres of the faces do not touch the Moon at its apogee when the Earth is at aphelion. Nor when the centres of the faces of the icosahedron are fitted to the aphelion of Venus do its vertices reach the Moon at its apogee when the Earth is at perihelion. This shows that there is some extra space between the perihelion of Mars and the vertices of the dodecahedron, as between the centres of the faces of the icosahedron and the aphelion of Venus. . . . I hope that I shall easily get moons of Mars and Venus into these spaces, Galileo, if you one day find such moons.' (KGW *4*, p. 310, l. 14 ff).

The two sections of the *Dissertatio* which concern Kepler's system of orbs and polyhedra are very short, and for any purpose other than our present one would be seen as rather insignificant parts of a work whose main interest lies elsewhere for the historian, as it probably did for Galileo. However, the passages relating to the 'secret' serve to confirm that the long years of arithmetical work on the orbit of Mars had not, apparently, altered Kepler's attitude to the quite different use of mathematics exemplified in the *Mysterium Cosmographicum*.

Kepler's annotations to the *Mysterium Cosmographicum*, 1621

The fact that Kepler allowed a second edition of the *Mysterium Cosmographicum* to be printed indicates that he still considered the work as being of some worth. However, his comment in the new dedicatory letter to the effect that all his later astronomical works might be seen as developments of chapters of the *Mysterium Cosmographicum*[16] may, at first glance, suggest that he considered the bulk of the earlier work to have been superseded by the later ones. It turns out that this impression is misleading (as is, indeed, suggested also by Kepler's references to the polyhedral theory in the *Dissertatio cum Nuncio Sidereo*). The problem Kepler set himself in the *Mysterium Cosmographicum* was a very large one: to explain 'the number, size and motions of the orbs' included almost all the astronomical or cosmological problems imaginable. That Kepler then proceeded to dissect this vast problem into several more specific ones, to each of which he assigned a chapter (or two), only confirms what we already know, namely that he was, even at the age of twenty-four, capable of considerable scientific insight. (He was also, as it turned out, capable of the hard work required to solve two of the most notable of the problems he had isolated.) As we have already seen, Kepler's notes of 1621 include references to other works which give a more satisfactory treatment of the matter in hand. However, a second glance at the new dedicatory letter assures us that Kepler does not say that *every* chapter of the *Mysterium Cosmographicum* has been superseded by a later, more successful, work. In fact, the theory which was the main subject of the book, the explanation of the number and spacing of the orbs by means of the Platonic solids, was never superseded by a theory put forward in any of his later works.

In the last chapter, Kepler's arguments in support of his theory were divided into three kinds: mathematical, astrological and numerological, and astronomical. The same division will be made in considering the annotations which appear in the 1621 edition of the *Mysterium Cosmographicum*, though in the present chapter little will be said about the first two kinds of argument, since it seems more appropriate to consider them in relation to the *Harmonice Mundi* (Linz, 1619), which is the subject of the next chapter.

(a) Mathematics

As we have seen in the last chapter, Kepler makes only two annotations to the specifically mathematical chapters of the *Mysterium Cosmographicum*, and both of these notes are quite trivial. Since there are chapters of the *Mysterium Cosmographicum* which are characterised as irrelevant (Chapter XI) or as now entirely out of date (Chapters XX and XXI), we may surely argue from the absence of such dismissive annotations that in 1621 Kepler still considered the mathematical chapters, Chapters III to VIII, as offering valid arguments in favour of his theory. We know, however, that he did have one very radical criticism of the mathematical reasoning to be found in these chapters: he no longer considered it appropriate to refer only to three-dimensional figures in seeking a mathematical explanation of the structure of the planetary system. This criticism is made in a note on the description of the system of orbs and polyhedra in Chapter II of the *Mysterium Cosmographicum* (KGW *8*, p. 50, note (1), referring to p. 46, l. 13). As we have seen, the style of this note is decidedly rhetorical, but the criticism is nonetheless entirely rational, its basis being summarised in the final sentence of the note: 'In constructing the number and sizes of the spheres, pure lines must be entirely left out; but in describing motions, which take place along lines, let us not despise lines and surfaces, which are the only origin of Harmonic proportions' (KGW *8*, p. 50, ll. 9–11). For our present purposes, it is sufficient to note that the first part of the sentence just quoted makes it entirely clear that Kepler's criticism is not designed to cast doubt upon the mathematical arguments adduced in the *Mysterium Cosmographicum*. The extension of the mathematical basis of the description of the planetary system, proposed in the second part of the sentence, is contained in the first book of *Harmonices Mundi Libri V* (Linz, 1619), as Kepler had pointed out earlier in his note (KGW *8*, p. 50, l. 3). This extension will be considered in the next chapter.

(b) Astrology and Numerology

Kepler's annotations to Chapters IX, XI and XII of the *Mysterium Cosmographicum* provide valuable evidence of the evolution of his astrological beliefs, and we shall have occasion to return to them in the next chapter, in connection with the astrological parts of the *Harmonice Mundi*. For our present purpose, however, the only part of this evidence that needs to be taken into account is the evidence that Kepler apparently no longer considered that astrological arguments had any important part to play in supporting his theory that the structure of the planetary system was derived from the Platonic solids.

Some of this evidence is very clear indeed. For example, the notes on Chapter IX are introduced with 'So this chapter is only an astrological game (*lusus astrologicus*) and should be thought of not as a part of the work but as a digression . . .' (KGW *8*, p. 59). Chapter XI is also characterised as irrelevant (KGW *8*, p. 62, l. 25). The remaining astrological chapter, Chapter XII, which is concerned with Aspects and the division of the Zodiac, is not directly described as irrelevant. However, almost all of the very extensive notes refer the reader to the much fuller discussion of the problem in *Harmonices Mundi* Books IV and V, and since neither of these books is in any way concerned with arguments in support of the theory described in the *Mysterium Cosmographicum* (though Book V, Chapter III does contain a summary of the theory), we must apparently take these numerous references as an indication that Chapter XII, also, now seemed to Kepler to be irrelevant to the main purpose of the *Mysterium Cosmographicum*. These notes on Chapters IX, XI and XII should not be seen as indicating that Kepler believed that astrological explanations had no part to play in his description of the planetary system. We know from many sources, not least from the *Harmonice Mundi*, that Kepler regarded astrology as being of considerable importance in this connection. Moreover, it is clear from all his astrological works that Kepler regarded astrology as having a firm basis in observation (see also Field, 1984c). However, the notes just cited clearly indicate that in 1621 Kepler considered that his discussion of the astrological powers of the planets should not be seen as relevant to the relation of the Platonic solids to the planetary orbs.

There are only two notes on the brief numerological chapter of the *Mysterium Cosmographicum*, Chapter X. One of the notes merely remarks that eight is not a factor of sixty (KGW *8*, p. 60, l. 32). The other points out that the origin of the nobility of numbers should be sought in geometry, and refers the reader to the first two books of the *Harmonice Mundi* for a more convincing explanation of their properties (KGW *8*, p. 60, l. 19 ff). This suggests that Kepler still regarded such considerations as relevant to his description of the planetary system but now thought his earlier treatment unsatisfactorily naïve; as well he might, since it consisted merely of identifying each number as being, for example, the number of solid angles in one of the regular bodies. In fact, the chapter is so phrased that it reads somewhat like a geometrical version of 'Green grow the Rushes O'.[17]

(c) Astronomy

Between the publication of the first edition of the *Mysterium Cosmographicum*, in 1596, and the publication of the second, in 1621, Kepler discovered the three astronomical laws which now bear his name. In *Timaeus*, Plato had made the first attempt to show the mathematical simplicity behind the misleading complexity of astronomical appearances, by separating the diurnal and annual motions of the Sun. With his first two laws Kepler had at last succeeded in establishing not only the mathematical simplicity of the planetary system but also its physical simplicity, in its relation to the Sun. His system did not, indeed, seem quite simple enough for Galileo's taste,[18] and Kepler apparently shared Galileo's attitude to the extent that he felt the need to explain the particular eccentricity of each ellipse. In this Kepler's physics and his metaphysics were at one: his physics told him that he must explain the deviation of each orbit from a circle (a circle which, moreover, was invoked again by the use of the 'mean radius' of the orbit in the third law) and his metaphysics required him to explain every apparently arbitrary element in his system.

The explanation of the ellipses is an important part of *Harmonices Mundi* Book V but it is (as we shall see in the next chapter) equivalent to explaining the thickness of each planetary sphere, and is therefore of no importance in connection with the theory described in the *Mysterium Cosmographicum*, which seeks to explain the spaces between the spheres.[19] For the purposes of the *Mysterium Cosmographicum*, the only significance of Kepler's first two laws was that they established that the orbits (or orbs) must be referred to the Sun, and allowed the actual paths of the planets to be specified with an unprecedented degree of accuracy. As we noted in the previous chapter, it is clear that in 1596 Kepler was ready to identify the geometrical centre of his system of orbs and polyhedra with what he regarded as the physical centre of the planetary system, namely the Sun. The first two laws provided observational justification for this identification. Their other contribution to Kepler's cosmological theory was to allow the calculation of more accurate orbits, and thus more accurate dimensions of the planetary spheres, against which the theory could be tested. Kepler presumably welcomed both contributions, though the latter was to lead him to make some modifications to his theory. (These modifications are discussed in Chapter V below.)

Kepler's annotations to the *Mysterium Cosmographicum* mention his third law only in connection with the relation between the dimensions of the orbs and their motions (that is, the periods of the planets). To Kepler, the order of priority of the properties of the planetary system seems to have been the one implied in his description of what his theory set out to explain, namely 'the number, size and motions of the orbs'. The third law is thus seen as giving a mathematical account of how the spacing of the planets determines their periods. In any case, the law refers to the mean distances of the planets from the Sun and thus has only an indirect connection with the distances between the planetary orbs, which are measured from the sphere containing the perihelion of one planet to the sphere containing the aphelion of the planet below it.

In fact, the treatment of the third law in *Harmonices Mundi* Book V is hardly less casual than its treatment in the notes to the *Mysterium Cosmographicum*, but it seems more appropriate to postpone discussion of the law until we come to consider the work in which it was first published. In the present chapter we shall concern ourselves only with the effect of the first two laws upon Kepler's belief in his theory that the structure of the planetary system was derived from the Platonic solids.

That the orbit of each planet is an ellipse with the Sun in one of its foci, and that the motion of the planet is such that a line joining the planet to the Sun sweeps out equal areas in equal intervals of time, are facts which provide a mathematical description of the planetary system that is very much simpler than any of those which preceded it. Moreover, the two laws suggest very strongly that any explanation of the form of the planetary system must be sought in the relation of the planets to the Sun. Perhaps Kepler might have taken this suggestion in a different manner if he had not already been a convinced Copernican (though it is hard to see how anyone but a convinced Copernican could have discovered the two laws). As it was, he was already convinced that the Sun must be seen as the physical 'centre' of the Universe, its importance being like that of God the Father in the Holy Trinity. The geometrical facts of the relation of the orbits to the Sun may therefore have seemed to be no more than a straightforward corollary to the physical facts. In its lengthy title the

Astronomia Nova is, after all, asserted to be concerned with celestial physics. The physics of the magnetic vortex which Kepler used to explain the orbits is qualitative rather than mathematical, but it seems clear that if Kepler had tried to give it a mathematical form he would have been looking for an explanation not of the ellipses themselves but of their deviations from circles centred on the Sun. It is, indeed, these deviations that he seeks to explain in physical terms in the *Epitome*, ascribing them to the changing geometrical relationship between the magnetic axis of the planet and the magnetic fibres emanating from the Sun.[20] A mathematical explanation of the deviations is given in *Harmonices Mundi* Book V: it consists of considering the ellipses, and thus the thicknesses of the planetary orbs, as being determined by mathematical musical harmonies, Ideas coeternal with the Creator which express themselves in His Creation in the same way as the Ideas of the Platonic solids expressed themselves by determining the spacings between the planetary orbs.

We might perhaps have expected that the discovery of the first two laws would tend to decrease the importance of the notion of a planetary orb. Such orbs are, however, one of the persistent traditional elements in Kepler's cosmology. Unfortunately, the very fact that they are a traditional element dispenses Kepler from giving a detailed definition of their status and properties. Although there can be no doubt that, at least by 1596, Kepler did not believe in solid orbs (see Field 1979b) it is clear from the *Mysterium Cosmographicum* that he does not use the word *orbis* in a way which would allow us to translate it as 'orbit', i.e. the planet's actual path in space. Instead it is clear that the words *orbis* and *sphaera* must signify a region of space through which the planet passes as it moves round the Sun. Moreover, this is not the same as a traditional 'sphere', namely the region which would be filled by a system of solid planetary orbs designed to move the planet in an appropriate orbit (see Chapter III). This impression that Kepler uses *orbis* or *sphaera* to signify the planet's 'own' region is confirmed by a brief reference to the subject in a letter he wrote in 1608: '. . . the sphere is either a body, in the opinion of physicists, or the three-dimensional region of space in which the planet may be found in the course of time'.[21] For Kepler, the function of these orbs seems to have been to define the geometrical structure of the Universe in terms of the most perfect of mathematical figures, the sphere: in a note added to Chapter II of the *Mysterium Cosmographicum* Kepler describes the Platonic solids as providing the archetypes for Creation and the orbs 'serving to construct the work' (KGW *8*, p. 50, note (3)). Apart from their role in its creation, the planetary orbs thus seem to have been an observed component of Kepler's Universe in rather the same way that the energy levels are an observed component of the Bohr atom.

Though the discovery of the first two laws did not apparently cause Kepler to change his conception of planetary orbs, it made it possible for him to calculate more accurate dimensions for the orbs. The extensive surviving manuscript material has allowed Bialas to describe in some detail how Kepler's work on planetary orbits progressed during the years which followed the publication of the *Astronomia Nova*. It appears that by the end of 1616 Kepler had calculated all the elements required to find the dimensions of the six planetary orbs.[22] Mutual gravitational effects among the planets cause a secular change in the elements of their orbits, and for some of the elements of some of the orbits the secular changes since Kepler's time are of the same order as the differences between Kepler's computed values and the values computed for 1600 on the basis of modern information (see Bialas, 1971, for references). Bialas therefore does Kepler the justice of comparing his values with modern 1600 values. Present day values (from Allen 1964) have been added merely for interest.

The elements with which we are concerned are the semi-major axis of the ellipse, a, and its eccentricity, e.

There are no longer any ambiguities regarding the correct choice of centres for the planetary orbs, and Kepler has succeeded in getting rid of epicycles, so it is clear that the inner radius of the planetary orb must be SA and the outer radius SA′ (see figure 4.1). We can see from the figure that

$$\text{SA} = a\,(1-e),$$
$$\text{and SA}' = a\,(1+e).$$

The thickness of the planetary sphere will thus be $2ae$. Table 4.3 shows Kepler's values for the semi-major axis and eccentricities of the planetary orbits, taken from Bialas (1971), and, in the last three columns, values derived from them.

Table 4.4 shows the outer and inner radii of the planetary orbs, first the 'observational' values (column 3) and then the values derived from the Platonic solids (column 4). Like the corresponding table in Chapter III, table 4.4 includes an anachronistic column of errors. Table 4.5 is a slightly modified version of the corresponding table in Chapter III, to allow an easy comparison of the values of 1621 with those of 1596.

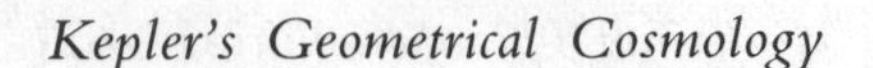

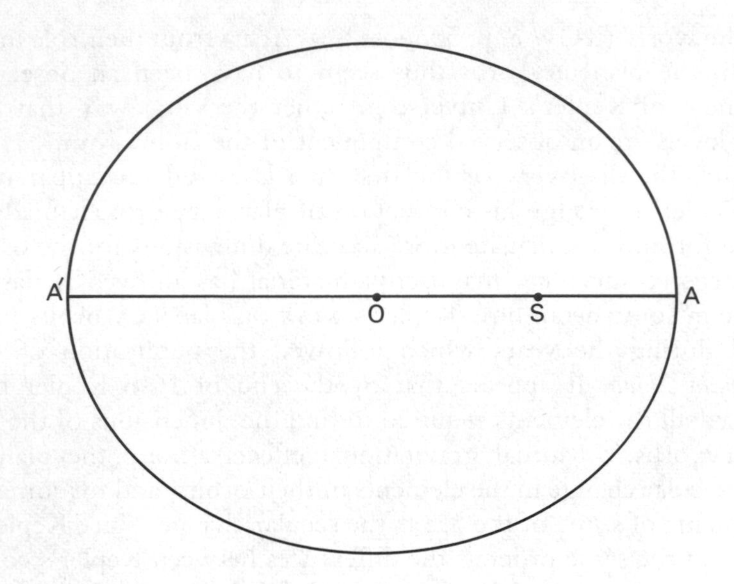

Figure 4.1 Standard ellipse
Centre O, focus S, apsides A and A′
Semi-major axis = a,
eccentricity = e.
$OA = a$
$OS = ae$, by definition,
$SA = a(1 - e)$
$SA' = a(1 + e)$.

Table 4.1 *Semi-major axis, a* (in AU), after Bialas (1971)

	Kepler	c.1600	c.1960
♄	9.510 00	9.554 73	9.538 84
♃	5.200 00	5.202 80	5.202 80
♂	1.523 50	1.523 68	1.523 69
⊕	1	1	1
♀	0.724 14	0.723 33	0.723 33
☿	0.388 06	0.387 10	0.387 10

Table 4.2 *Eccentricity, e,* after Bialas (1971).

	Kepler	c.1600	c.1960
♄	0.057 00	0.056 81	0.055 65
♃	0.048 22	0.047 84	0.048 45
♂	0.092 65	0.093 02	0.093 38
⊕	0.018 00	0.016 88	0.016 72
♀	0.006 92	0.006 98	0.006 79
☿	0.210 01	0.205 56	0.205 63

Table 4.3 *Calculating dimensions of the orbs*

	a	*e*	*ae*	*a* + *ae*	*a* − *ae*
♄	9.510 00	0.057 00	.54207	10.052	8.9679
♃	5.200 00	0.048 22	.25074	5.4507	4.9493
♂	1.523 50	0.092 65	.14115	1.6647	1.3825
⊕	1	0.018 00	.01800	1.0180	0.9820
♀	0.724 14	0.006 92	.00501	0.72915	0.71913
☿	0.388 06	0.210 01	.08150	0.46956	0.30656

The values in table 4.4 take no account of the orbit of the Moon, but, since the table shows that the inner surface of the orb of Mars and an inscribed dodecahedron are predicting too large a radius for the outer surface of the orb of the Earth, the orb of the Moon would clearly act to improve the agreement between theory and observation. The same is true for the prediction of the outer surface of the orb of Venus, whose radius is also too large, and would be decreased if the orb of the Moon were taken into account.

Table 4.4 also seems to show what Kepler meant by his remark near the end of the *Dissertatio cum Nuncio Sidereo* (Prague, 1610) that when he had recently recalculated the orbits of Mars, the Earth and Venus from Tycho's observations he had found that the spaces between their orbs were slightly larger than those required by his theory (see above and KGW *4*, p. 310, l. 14 ff). This suggests that Kepler must have obtained some values for *a* and *e* for the orbit of Venus rather earlier than the date indicated in Bialas's table (see our table 4.6). As Bialas remarks in the paragraph above the table: extensive as they are, the manuscripts do not provide a complete diary of Kepler's work on the orbits (Bialas 1971, p. 127).

The figures shown in table 4.4 are not displayed in the new edition of the *Mysterium Cosmographicum*. The comparison between the theoretical and 'observational' values of the radii of the spheres, in Chapter XIV (where the centre of the system is the centre of the Great Orb), has only four annotations in the second edition of the work. Three are apparently designed merely to clarify the expression of certain points. The fourth discusses the problem of the relative distances of the Sun and Moon from the Earth (KGW *8*, p. 84, l. 30). It is clear from this note that in 1621 Kepler believed that his earlier figures had made the orb of the Moon larger than it should have been, but he does not discuss any significance this might have in assessing the relation of the planetary orbs to the Platonic solids. The only

Table 4.4 *Orbs and polyhedra, 1621.* Values for Mercury use midsphere.

		observed distance	theoretical distance	error (th − obs)/obs%
♄	aph	10.0521	10.6730	
	peri	8.9679	8.8281	− 2
♃	aph	5.4507	5.0969	
	peri	4.9493	4.6280	− 6
♂	aph	1.6646	1.5427	
	peri	1.3823	1.2811	− 7
⊕	aph	1.0180	1.0180	0, by definition
	peri	0.9820	0.9820	
☿	aph	.72915	.78035	+ 7
	peri	.71913	.76963	
♀	aph	.46955	.54421	+16
	peri	.30656	.50247	

Table 4.5 *Orbs and polyhedra, 1596, heliocentric, with corrections from Aiton (1981).* Values for Mercury use midsphere.

		observed distance	theoretical distance	error (th − obs)/obs%
♄	aph	9.727	10.011	+ 3
	peri	8.602	8.854	
♃	aph	5.492	5.109	− 7
	peri	4.999	4.650	
♂	aph	1.648	1.550	− 6
	peri	1.393	1.310	
⊕	aph	1.042	1.042	0, by definition
	peri	0.958	0.958	
♀	aph	0.721	0.762	+ 6
	peri	0.717	0.757	
☿	aph	0.481	0.535	+11
	peri	0.233	0.260	

Table 4.6 *The progress of Kepler's work on the planetary orbits*, after Bialas (1971, p. 127)

Date	Mercury	Venus	Sun (and Earth)	Mars	Jupiter	Saturn
1600 July				ecc.		
1601 Nov				circle		
1603 Sept				oval		
1605 May				*a*, *e*		
1609 Nov	aph., *e*					
1614 Jan				epochs		
1614 Summer		aph., *e* *a*, *i*, ☊	Tab.aeq.			
1614 Dec	aph., *e*					
1615 March	*e*					
before 1616					aph., *e*	
1616					*a*, *e*,	*a*, *e*
1616 Sept					*a*, *e*	
1616 Oct	aph., *e* *a*, ☊					
1617	epochs					
1622 Summer					epochs	
1622 Autumn			epochs	epochs	epochs	epochs
1624 Summer	epochs	epochs	epochs	epochs	epochs	epochs, *a*
1624 Summer		epochs			epochs	aeq.sec.
1624 Dec	epochs					

Key: *a* semi-major axis
aeq.sec. change in mean
aph. aphelion motion

e eccentricity
ecc. circle eccentric circle
epochs epochs of perihelion passage etc.

i inclination of orbit
☊ ascending node and direction of line of nodes

oval oval orbit (for Mars)

Tab.aeq. Table of equations for *e*.

annotation to Chapter XV (where the centre of the system is transferred to the Sun) refers the reader to the new description of the planetary system in the *Astronomia Nova* (KGW *8*, p. 90, 1. 2) but does not mention that the dimensions of the new orbits are different from those given in the *Mysterium Cosmographicum*. It thus appears that Kepler wishes to abide by his conclusion that the radii deduced from his theory agree quite well with the radii deduced from observation.

In the first edition of the *Mysterium Cosmographicum*, Kepler was able to explain away the discordances between theory and observation by pointing to the uncertainties in the 'observed' orbits. The

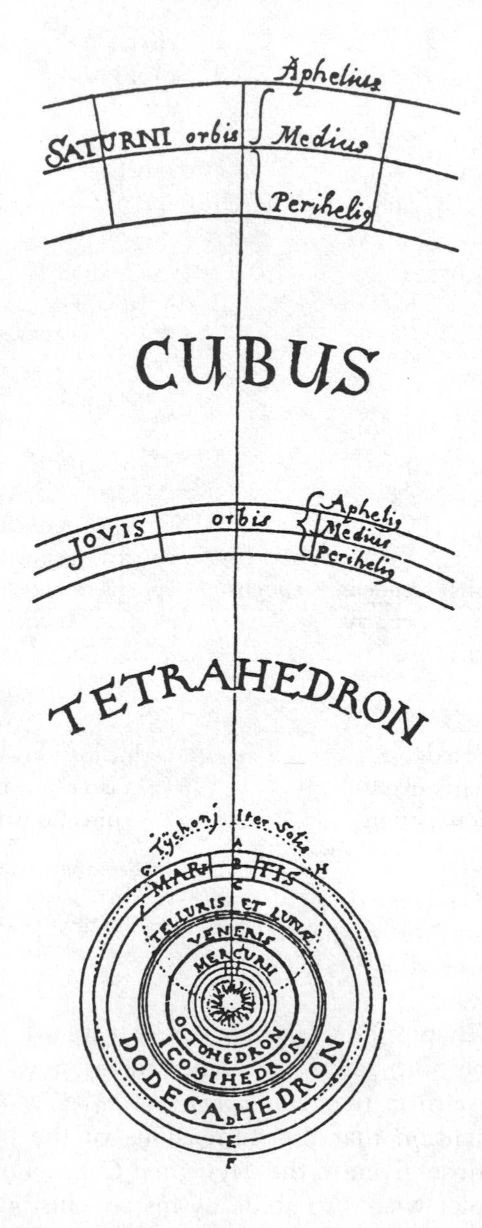

Figure 4.2 Orbs and polyhedra (*Harmonices Mundi* Book V, Ch. III)

continued discordance must therefore have appeared to shed doubt on either the new orbits or the old theory. Kepler cannot have known, as we do, that his new orbits were very accurate indeed, but it is nevertheless clear from his annotations to the later chapters of the *Mysterium Cosmographicum* that he believed that it was the theory that must be modified. The notes repeatedly refer the reader to *Harmonices Mundi* Book V for an account of the modifications.

Although Kepler's notes on the *Mysterium Cosmographicum* do not give a full discussion of how the agreement between theory and observation has been affected by the new orbits, some kind of assessment of Kepler's opinions on the matter might well be extracted from what he says, and it is unlikely that the absence of an explicit discussion should be interpreted as indicating that he wished to ignore the problem. It is much more probable that Kepler's vagueness is to be explained by the fact that by the time he came to write his notes, in the last two weeks of June 1621 (according to Hammer, KGW *8*, *Mysterium Cosmographicum* Afterword), he had already published two further descriptions of his theory, both of which discuss its agreement with observation. One of these is in *Harmonices Mundi* Book V (Linz, 1619) and the other is in Book IV of the *Epitome* (Linz, 1620). The two descriptions seem to be entirely consistent with one another and we shall concentrate on the former mainly because it contains a considerably fuller discussion of the agreement with observation.

In *Harmonices Mundi* Book V, Kepler uses the mathematical and musical results of Books I, II and III to explain the observed structure of the planetary system. Chapters I and II of Book V contain discussions of the five regular solids and their relation to the ratios found in musical theory. (These chapters of Kepler's work will be discussed in Chapter V below.) Chapter III of Book V contains a summary of the astronomy required for a study of celestial harmonies. The relation between the five Platonic solids and the six planetary orbs is first introduced to explain the number of the planets (KGW *6*, p. 298, l. 18). The account of the theory consists of little more than a diagram (see figure 4.2) and a reference to the *Mysterium Cosmographicum*. The next section is concerned with the sizes of the orbs:

> 'Fourthly, as for the proportion between the planetary orbs, the proportion between two neighbouring orbs is always such (as

> may easily be seen) that each one of these proportions is close to the proportion between the orbs associated with a particular one of these five solid figures, that is, it is close to the proportion between the circumscribed and inscribed spheres of the solid figure. But it is not exactly (*plane*) equal to it, as I once dared to promise from perfected astronomy. For, having calculated the distances from Brahe's observations, I found that, if the angles of the Cube were placed on the inner circle of Saturn, the centres of its faces almost touched the mean circle of Jupiter . . .' (KGW *6*, p. 298, l. 33 ff).

Kepler proceeds to work in through the system of orbs and polyhedra, noting the discrepancies, but not giving any actual numerical values. (The arithmetic sets in only in Chapter IV.) The following paragraph sums up the situation:

> 'From this it appears that the exact proportions of the distances of the planets from the Sun were not taken from the five regular solid figures alone; for the Creator does not depart from his Archetype (*non aberrat enim ab Archetypo suo Creator*), the Creator being the true source of Geometry and, as Plato wrote, always engaged in the practice of Geometry' (KGW *6*, p. 299, l. 32).

The rest of *Harmonices Mundi* Book V is concerned with explaining the subtleties of the actual Archetype, which involved harmonic ratios as well as the Platonic solids.

Kepler's belief in his system of orbs and polyhedra, 1621

As we have seen, the 'perfected astronomy' derived from Tycho's observations led Kepler to modify the theory described in the *Mysterium Cosmographicum*. However, there was no need for any drastic modification of the theory, which Kepler clearly still regarded as intellectually satisfying, mathematically justified and in fairly good agreement with the observations. He believed that the differences between the theory and the observations were real and must be explained, but they did not seem to him to be so large as to cast doubt upon the essential correctness of the theory. Moreover, the modified explanation of the orbs, which is described in *Harmonices Mundi Libri V*, had a wider mathematical basis than the earlier explanation, being derived from the relations of polygons to the circle and the relations of polygons among themselves. It also made a much more sophisticated attempt to explain the 'known' properties of astrological Aspects. As might have been expected, Kepler regarded the later theory as a considerable advance upon the earlier one.

V
Harmonices Mundi Libri V

One of the notes on the title page in the second edition of the *Mysterium Cosmographicum* tells us that when the first edition was published Kepler intended to write further works in the same vein. The exactitude of this recollection of 1621 is confirmed by several letters written in the late 1590s. When writing to Herwart von Hohenburg in March 1598 Kepler even went so far as to give titles for the projected works: 'I intended to write four short books on cosmography, 1. On the universe, and particularly the parts of the world that are at rest . . . 2. On the parts that move . . . 3. On individual globes, particularly that of the Earth . . . 4. On the relation between the heavens and the Earth . . .'[1] It is clear from the note in the second edition of the *Mysterium Cosmographicum* that Kepler had hoped that his four new works would be similar in character to the *Mysterium Cosmographicum*, that is, that they would give accounts of their subjects deduced a priori from mathematical principles. He gives as his reason for abandoning the project of writing these books his realisation that 'the heavens, the first of God's works, were laid out much more beautifully than the remaining small and common things'.[2] 'So', Kepler adds, 'the Prodromus [i.e. the *Mysterium Cosmographicum*] was a work on its own (*egregius*): for it was not followed by any *Epidromus* [i.e. sequel] such as I had proposed to write . . . However, the reader could regard my astronomical works, particularly my books on Harmonics, as the true and proper sequel to this work'.[3]

A first project for what was later to become the 'books on Harmonics' was, in fact, formed before Kepler had entirely abandoned his intention of writing the four cosmographic works, for he wrote to Maestlin in August 1599: 'If I were in Tübingen I would be planning to write a new short book. So it is, indeed, quite true what Maestlin said to people like me – that they cannot do anything without him. For I shall not deal with the four books on cosmography unless I survive Tycho's editions. The title of the book

I am planning to write is *De harmonia mundi* . . .'.[4] The following three hundred and thirty-two lines of the letter give a summary of the material that is to be included in *De harmonia mundi*. The correspondence with the actual contents of *Harmonices Mundi Libri V* (Linz, 1619) is very close. In particular, the summary of 1599 includes two of the most important elements of the published work: 'musical' ratios are deduced by means of geometry, and these ratios are then identified with ratios between velocities of planets (see, for example, the table at line 380 of the letter, KGW *14*, p. 52). In the letter to Maestlin, Kepler presents his ideas in a rather informal manner. For example, he uses geometrical results as and when he requires them, as if they were too well known to stand in need of proof – as they may well have been to Maestlin. However, when writing to Herwart von Hohenburg several months later he gives a very brief but much more formal outline of the work he has in mind:

> '. . . I have drawn up a plan and first sketch of a short book, a cosmographic dissertation that I shall call *De Harmonice Mundi*. There will be five books or chapters.
> 1. Geometrical, on constructible figures.
> 2. Arithmetical, on solid ratios
> 3. Musical, on the causes of harmonies
> 4. Astrological, on the causes of Aspects
> 5. Astronomical, on the causes of the periodic motions'.[5]

At first glance, this plan appears to be designed to demonstrate that there is a unity among the traditional subjects of the quadrivium, but the resemblance to the plan of *Harmonices Mundi Libri V* is in fact very close. The main differences seem to be that the projected first book became *Harmonices Mundi* Books I and II, while the projected third book was absorbed into the eventual Book III as an introduction and a first chapter which discuss some properties of numbers and give geometrical justifications for assigning greater significance to some particular arithmetical ratios. We may note, also, that *Harmonices Mundi* Book V explains rather more than was promised in the early sketch, since it not only accounts for the periods of the planets but also for the dimensions and eccentricities of their orbits.

Nearly twenty years were to elapse between these sketches and the publication of *Harmonices Mundi Libri V*. However, the surviving correspondence proves that throughout these years Kepler repeatedly returned to the idea of a universal mathematical harmony. The evidence thus seems to justify the historian in availing himself of

Kepler's permission to regard the *Harmonice Mundi* as in some sense a sequel to the *Mysterium Cosmographicum*. The later work is not, however, a sequel in the sense that it is concerned with any further defence of the theory put forward in the earlier one to account for the number and spacing of the planetary orbs. As we have seen in Chapter IV, Kepler considered his theory was not in need of further defence, being substantially correct and, moreover, not under attack. The purpose of the *Harmonice Mundi* is to put forward a more comprehensive cosmological theory, derived from geometry and containing a slightly more sophisticated version of the theory described and defended in the *Mysterium Cosmographicum*. The new theory has a more elaborate mathematical foundation and is wider in scope than its predecessor, providing explanations of musical, astrological and astronomical facts of which the earlier theory took no cognizance, or which it could not adequately explain.

The plan of *Harmonices Mundi Libri V*

The 'short book' planned in 1599 was published twenty years later as a folio of about three hundred and twenty pages. Of the five books which make up the work, the first two deal with geometry, and the third with musical theory (in terms of the properties of certain arithmetical ratios), while the two final books contain applications of the mathematical and musical theorems to construct explanations for the efficacy of Aspects, in Book IV, and the structure of the Solar system, in Book V (it being taken as an accepted fact that the five Platonic solids explain the number of the planets and the approximate sizes of the spaces between their orbits).

The order in which the books are presented may at first seem to have been dictated by the necessity of proving theorems before using them, but closer scrutiny reveals that this is not entirely the case: the astrological book, Book IV, uses the geometrical results directly, not in the musical form which they have been given in Book III, as Kepler himself remarks (HM IV, Ch. IV, KGW *6*, p. 234, l. 32), and the musical results of Book III are not applied until Book V. It is possible that the central position of the book concerned with music seemed appropriate in a work on Harmony, but it should be noted that Kepler, like other writers on *musica mundana*, uses the word 'harmony' and its cognates in a much wider sense than the purely musical one. For example, the title of Book IV refers to 'harmonious configurations of stellar rays' when, as we have seen, the harmony is, according to Kepler's explicit statement, not a musical one. The position of the musical book immediately after the geometrical ones was probably designed to emphasise that the 'musical' ratios, long seen as arithmetical in origin, had now been given a basis in geometry. This basis helps to justify the wide sense which Kepler apparently gives to 'harmony' in the title of his work as well as in its text. It also allows him to see the musical ratios among the velocities of the planets as a consequence of the fact that God is a Platonic Geometer, whereas they might otherwise perhaps have suggested that He was something of a Pythagorean Numerologist.[6]

Geometry: *Harmonices Mundi* Books I and II

In *Harmonices Mundi* Book I Kepler deduces the geometrical results which will be used in Book III to show that particular significance must be attached to certain ratios of small integers. He makes no claim that the work in Book I is mathematically original. Indeed, he goes so far as to say that if Proclus had left a commentary on Book X of Euclid's *Elements* then he, Kepler, would not have needed to write the present work (HM I, Introduction, KGW *6*, p. 15, l. 16). The connection with Proclus has, in fact, already been made, by a quotation from his *Commentary on the First Book of Euclid's Elements*, which appears on the title page of *Harmonices Mundi* Book I, in the original Greek (see figure 5.1). A slightly longer part of the same passage appears, in Latin translation, on the title page of *Harmonices Mundi* Book IV. In Morrow's translation, the passage quoted as an epigraph to Book I reads

> 'Mathematics also makes contributions of the very greatest value to physical science [i.e. the study of Nature]. It reveals the orderliness of the ratios according to which the Universe is constructed and the proportion that binds things together in the cosmos, . . . It exhibits the simple and primal causal elements as everywhere clinging fast one to another in symmetry and equality, the properties through which the whole heaven was perfected when it took upon itself the figures appropriate to its particular region'.[7]

Proclus seems to be describing a use of mathematics like that in Plato's *Timaeus*, where stress is laid upon considerations of symmetry and simplicity which are as much aesthetic as mathematical. As we have seen in Chapter III, the geometrical

IO. KEPLERI
HARMONICES MUNDI
LIBER I.

DE FIGVRARVM REGVLA-
RIUM, QUÆ PROPORTIONES HAR-
monicas pariunt, ortu, claſsibus, or-
dine & differentijs, causâ ſcientiæ
& Demonſtrationis.

PROCLUS DIADOCHUS
Libro I. Comment. in I. Euclidis.

Πρὸς δὲ τὴν φυσικὴν θεωρίαν (ἡ μαθηματικὴ) τὰ μέγιστα
συμβάλλεται, τήν τε τῶν λόγων εὐταξίαν ἀναφαίνουσα, καθ' ἣν
δεδημιούργηται τὸ ΠΑ͂Ν, &c: καὶ τὰ ἁπλᾶ καὶ πρωτουργὰ στοι-
χεῖα, καὶ πάντῃ τῇ συμμετρίᾳ καὶ τῇ ἰσότητι συνεχόμενα δείξα-
σα, δι' ὧν καὶ ὁ πᾶς οὐρανὸς ἐτελειώθη, σχήματα τὰ προσ-
ήκοντα, κατὰ τὰς ἑαυτοῦ μερίδας ὑποδε-
ξάμενος.

Cum S. C. M^tis. Pri-

vilegio ad annos XV.

LINCII AUSTRIÆ
Excudebat Johannes Plancus,
ANNO M. DC. XIX.

Figure 5.1 Title page of *Harmonices Mundi* Book I, 1619.

arguments in Kepler's *Mysterium Cosmographicum* were presented in the same way, as arrays of facts rather than lines of reasoning. However, the geometry in *Harmonices Mundi* Books I and II is of a very different order. The books are constructed as series of axioms, definitions and propositions, reminiscent of the *Elements* not only in their style but also in their mathematical rigour.

As we have seen, Kepler implied in his introduction to *Harmonices Mundi* Book I that the work should be seen as closely related to Book X of the *Elements*. It seems probable that this remark is meant to apply to the whole of the *Harmonice Mundi*, since the Introduction ranges very widely and only its final paragraph, with the marginal note 'the purpose of this first book', seems to serve as a specific introduction to the geometrical work of Book I (KGW *6*, p. 19, l. 21). However, even the geometry in Book I amounts to something rather more than a commentary on Euclid's work. Kepler is concerned with the relation of the side of a regular polygon to the diameter of the circle in which it is inscribed, whereas Euclid's Book X is concerned with the classification of different types of surd. From Euclid's point of view, of course, Book X is concerned with the classification of lines of certain magnitudes which are or are not commensurable with some given magnitude, in some degree. It is well nigh impossible, in giving a description, to avoid using algebraic terminology which serves to emphasise the divergence between Euclid's work and Kepler's. What Kepler has done is to reorganise Euclid's results in such a way that they can be used to classify regular polygons according to the degree of commensurability of their sides with the diameter of the circle in which they are inscribed. This classification of polygons is original to Kepler and is not to be found in Euclid.

However, a hint of its possibility may perhaps have been taken from the *Elements*, for Proposition 11 of Book XIII is 'If in a circle which has its diameter rational an equilateral pentagon be inscribed, the side of the pentagon is the irrational straight line called minor',[8] and the following proposition is also concerned with the relation between the circle and an inscribed polygon: 'If an equilateral triangle be inscribed in a circle, the square on the side of the triangle is triple of the square on the radius of the circle' (Euclid trans Heath, 1956, vol. III, p. 466). Furthermore, it should be noted that the comparisons of the radii of incircles and circumcircles, which Kepler describes in the Preface to the *Mysterium Cosmographicum* (KGW *1*, p. 12, see Chapter III), involve drawing diagrams and solving triangles which make a comparison of the side of the polygon with the radius of the circumcircle almost inevitable, since the side of the polygon forms the third side of the triangle whose other two sides are the radii of the incircle and of the circumcircle (see Appendix 3).

In *Harmonices Mundi* Book IV, when Kepler comes to apply the musical relations which he has derived from the geometrical work of Book I, he justifies this application by drawing an analogy between the circle and the human soul (HM IV, Ch. I. KGW *6*, p. 224, ll. 20–30). In view of this alleged analogy, it may be as well to point out that there are very good mathematical reasons for looking to the circle when investigating the properties of regular polygons. To put it in modern terms: since a regular *n*-gon has *n*-fold rotational symmetry it is clear that a circle may be drawn through all its vertices or to touch it at the mid-points of all its sides. It is natural to use these two circles as aids to constructing regular polygons. Euclid is not given to appeals to symmetry, but when regular polygons are introduced, in Book IV of the *Elements*, each one is, indeed, constructed first by circumscription about a circle and then by inscription within a circle. Kepler, who is very interested in symmetry, explicitly proves, in the fourth section of *Harmonices Mundi* Book I, that 'All regular figures can be placed in a circle so that each of their angles lies on it' (HM I, sect. IV, KGW *6*, p. 21, l. 2). The proof, as Kepler points out, depends on propositions 21 and 24 of Book III of the *Elements*.

Harmonices Mundi Book I begins with a definition of the term 'regular' (*regularis*) applied to plane figures, since Euclid uses the phrase 'equilateral and equiangular'. More preliminary definitions and propositions follow, the most significant being the definition of 'to know' (*scire*) in section VII: 'To know in geometry is to measure in terms of some known measure. In this matter of inscribing figures in circles the known quantity is the diameter of the circle.' It is in terms of this 'knowledge' that Kepler proceeds to construct his classes of magnitudes.

Sections XII to XXIX contain the definitions of these classes, with some associated propositions. The first class is that of the magnitude which is equal to the measure (sect. XII, KGW *6*, p. 22). The second comprises what a modern mathematician would call 'rational' magnitudes; Kepler, making a literal translation of Euclid's adjective (ῥητη) prefers to call them 'expressible' (*effabilis*) (sect. XIII). The third class comprises lengths whose squares are commensurable with the square of the diameter, lengths 'expressible in power' (sect.

XIV). The members of the following, unnumbered, classes are all 'inexpressible'. Kepler then implicitly explains his choice of the term 'expressible' and its cognates by complaining that the usual Latin term for 'inexpressible' magnitudes, namely 'irrational', carries with it 'the danger of much ambiguity and absurdity' (sect. XV). It is in this section of Book I that Euclid's system of surds begins to appear, at first merely by the use of their names, but later in the form of numerous references to the propositions of Book X of the *Elements*. The rewriting of Euclid continues until section XXIX. In his introduction Kepler had taken the precaution of warning prospective readers that if they were not very thoroughly versed in mathematics they might prefer to begin the book at section XXX (HM I, Introduction, KGW *6*, p. 20, l. 7). Caspar's last note to this first part of Book I ends with a thought that is only a hair's breadth away from being reprehensibly unhistorical but must surely occur to many a modern reader: 'It is amazing how Kepler manages to describe these irrational magnitudes purely geometrically' (KGW *6*, p. 522).

In section XXX polygons reappear. They are assessed singly or in groups in the series of propositions that follows, in an order determined by their geometrical properties. The treatment of the square is a simple, but otherwise typical, example: 'The side of the Tetragon can be constructed geometrically, using the angles, outside the circle, and if it is inscribed in a circle the side belongs to the third class of knowledge, and its square to the second, as does the area of the figure . . .' (HM I, sect. XXXV, KGW *6*, p. 36). Kepler then proceeds to discuss the octagon and the star octagon, which can be constructed from the square. The diagram in this section, like those in many of the following ones, seems to aspire to the nature of a rectilinear design for an elaborate rose window (see figure 5.2).

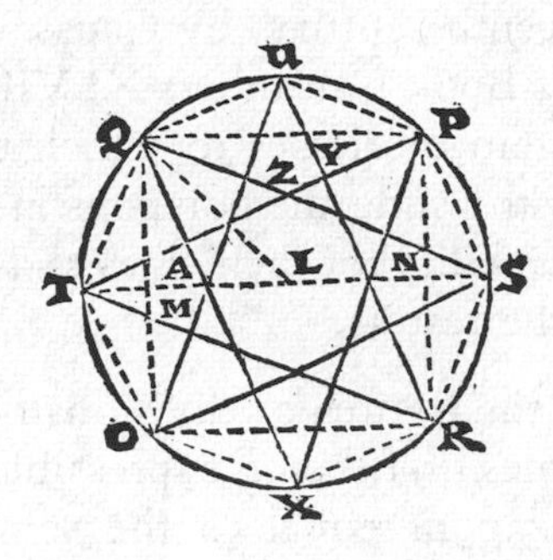

Figure 5.2 Square and octagons in a circle (HM I, sect. XXXVI)

Book IV of the *Elements* considers the construction of triangles, a square, a regular pentagon, a regular hexagon and a regular pentekaedecagon (15–gon). *Harmonices Mundi* Book I is more comprehensive: Kepler considers a diameter (XXXIV), a square (XXXV), an octagon and a star octagon (XXXVI), a hekkaedecagon (16–gon, XXXVII), a triangle and a hexagon (XXXVIII), a dodecagon and a star dodecagon (XXXIX), a 24–gon and the figures obtained by repeated doubling of that number of sides (XL), a decagon and a star decagon (XLI), a pentagon and a star pentagon (XLII), an icosigon (XLIII), and a pentekaedecagon and star pentekaedecagons (XLIV). The inclusion of the diameter as the first regular polygon is apparently original to Kepler. It is an interesting mathematical insight, and one that has found significant applications in the twentieth century (see, for example, Coxeter, Longuet-Higgins and Miller, 1953), but we cannot be sure that Kepler's purpose in introducing it was purely mathematical, particularly since he does not in fact attempt to describe the diameter as a degenerate polygon. He was to use the diameter in deducing harmonic ratios in Book III, and it is more than possible that an awareness of this application played a part in determining the mathematical content of Book I.

Sections XLV to XLVII (KGW *6*, pp. 47–62) deal with proofs that other regular polygons cannot be inscribed in a circle 'by geometrical means' i.e. using a straight edge and compasses. Section XLV is chiefly concerned with the regular heptagon. After having shown that 'geometrical means' are of no avail, Kepler turns to algebra, remarking that this treatment of the heptagon is due to Jobst Bürgi (KGW *6*, p. 50, l. 24). Some elements in Kepler's proof appear to be questionable (see Caspar's note, KGW *6*, p. 526). A correct proof was provided in the nineteenth century, by Gauss.

The final sections of Book I, numbers XLVIII to L, are concerned with defining the system of classes to which the regular polygons will be assigned and with listing the polygons in the order determined by the degree of 'knowability' of their sides and their areas. The very last section sums up the results:

'L. Comparison of the figures or divisions of the circle

The diameter comes first, being expressible in length. Second is the side of the hexagon, equal to the semidiameter, and thus expressible in length. In the third place are the tetragon and the trigon, because they have sides expressible only in power. In the fourth rank are the sides of the dodecagon and the decagon and their associated stars . . .

Further to these properties of the side there is another indication of rank, since the figures differ in the aptness and perfection of the area the figure encloses. In this, after the diameter (which has no area, merely dividing the area of the circle into two equal parts, as Ptolemy points out, just as it divides the circumference), the leading position is that of the Tetragon and dodecagon, which have expressible areas . . .' (HM I, sect. L, KGW *6*, p. 63, l. 26 ff and p. 64, l. 4 ff).

The ordering of the polygons thus corresponds to Euclid's ordering of the surds that describe their sides or their areas: for example, binomials and apotomes take precedence over majors and minors. Euclid's ordering was equivalent to listing the surds in order of increasing algebraic complexity, a purely mathematical affair. Kepler, however, is concerned with placing polygons in order of their rank, the closeness of the relation between the side of the figure and the diameter of the circle being taken as an indication of simplicity and therefore of nobility. In later books of the *Harmonice Mundi* this nobility will be taken as an indication of the figure's power to contribute to an archetype. Book I ends with this classification of polygons.

Harmonices Mundi Book II is concerned with establishing a different classification of regular polygons, by considering their 'sociability', that is their capacity to combine with polygons of the same or different kinds so as to form a 'congruence', that is a flat pattern which will cover the plane (a tessellation) or a polyhedron whose vertices are all alike (a uniform polyhedron). The concluding sections of Book II are thus very similar in style to the concluding section of Book I, but the main body of the book is of a very different order.

Although tessellation patterns appear in several earlier works, for example in Dürer's *Underweysung der Messung mit Zirkel und Richtscheyt* (Nuremberg, 1525), Kepler is the first writer to deal systematically with the problem of constructing all the tessellations formed by regular polygons. His interest in the problem appears to date back at least as far as 1599, when he discussed tessellations in a letter he wrote to Herwart von Hohenburg.[9] The work on tessellations in *Harmonices Mundi* Book II is of interest not only to historians of mathematics but also, it seems, to some modern mathematicians, who have developed generalisations of Kepler's work (see Grünbaum & Shephard, 1977). From flat patterns of polygons Kepler turns to polyhedra, considered as constructed from the polygons which form their faces. This conception of polyhedra is the one Plato used in *Timaeus* when constructing his solids from the basic triangles, but Kepler apparently regarded it more or less as a mathematical convenience which allowed him to develop his laborious, but rigorous, proofs by exhaustion (see Field 1979a).

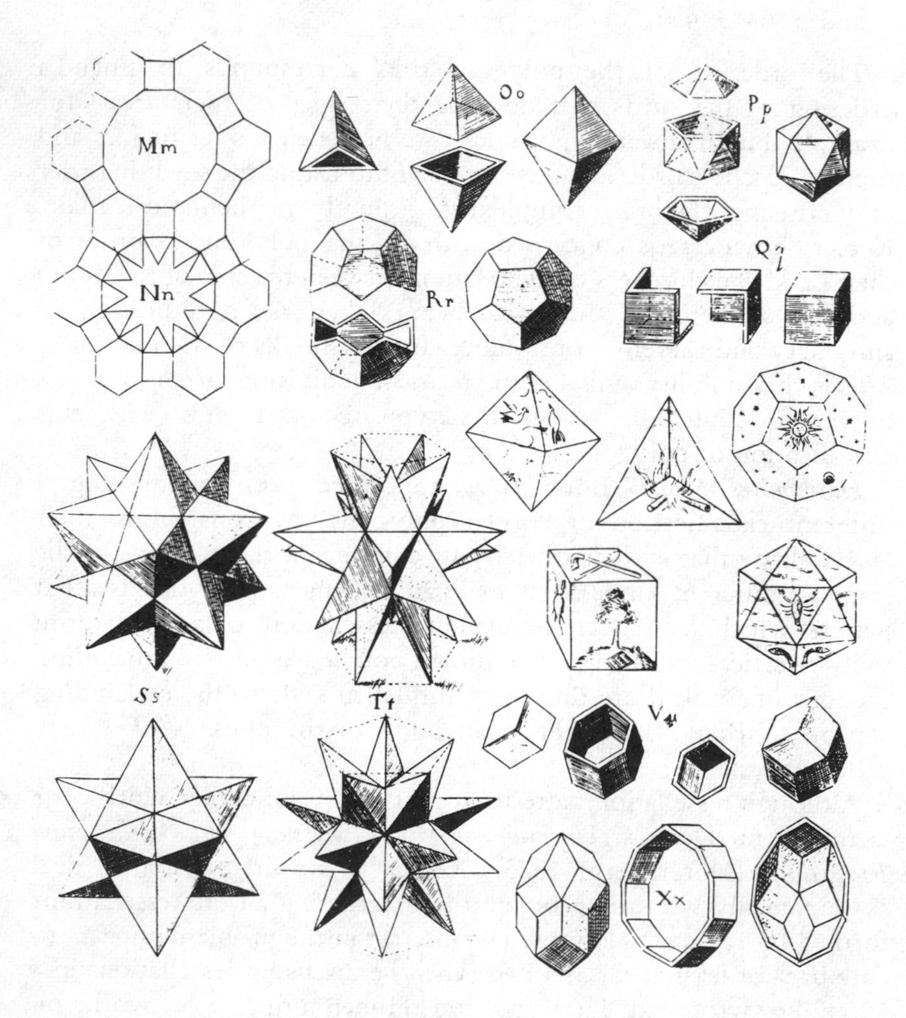

Figure 5.3 Tessellations and polyhedra (plate from HM II)

Harmonices Mundi Book II contains Kepler's first printed account of the two new regular polyhedra which he had discovered (HM II, sect. XXVI, KGW *6*, p. 82). Each of these figures has twelve star pentagon faces whose central parts are hidden inside the polyhedron (see figure 5.3, Ss and Tt). This concept of a face which cuts into the solid so that the face is not entirely visible on the outside of the body appears to be original to Kepler. His description of the new polyhedra points out that their structure is analogous to that of the star pentagon, whose sides cut through its area in the same way as the faces of the new polyhedra cut through the bodies of the solids. Kepler may well have discovered the two new regular polyhedra by considering the effect of converting the pentagons found in the Platonic dodecahedron and icosahedron into the corresponding star pentagons (see Field 1979a). In any case, the analogy with the star pentagon seemed significant to Kepler and he therefore gave the new polyhedra the same status in relation to the corresponding Platonic solids as that he had given to the star pentagon in relation to the convex pentagon (HM I, sect. II, KGW *6*, p. 20), that is, he regarded them as secondary figures, mere derivatives of the Platonic dodecahedron and icosahedron. Their inferior status is emphasised by the fact that the new solids are not described in the section which deals with the Platonic solids, but in the following section, which also considers two incomplete polyhedra constructed from star polygons. It is, however, clear that Kepler realised that the two star polyhedra were regular, that is, that their faces were all the same and met in the same way at each vertex of the solid.

Having dealt with solid congruences which each involve only one kind of polygon, Kepler turns to solids whose faces are of more than one kind, and proceeds to give a lengthy proof by exhaustion that there are exactly thirteen convex uniform polyhedra, namely the Archimedean solids. Historians seem to be agreed that Kepler was the first mathematician to prove that there were exactly thirteen such solids (at least in modern times), and that his is the earliest surviving description of their structure. Pappus had merely listed the faces of the solids and the work of Piero della Francesca and Luca Pacioli, published by Pacioli in *De Divina Proportione* (Venice, 1509), only described six of the thirteen solids. Kepler's work on the Archimedean solids must therefore be seen as historically significant. The fact that he does not present it as being of any particular importance must presumably be seen as indicating that for Kepler the real interest of his work lay elsewhere: the

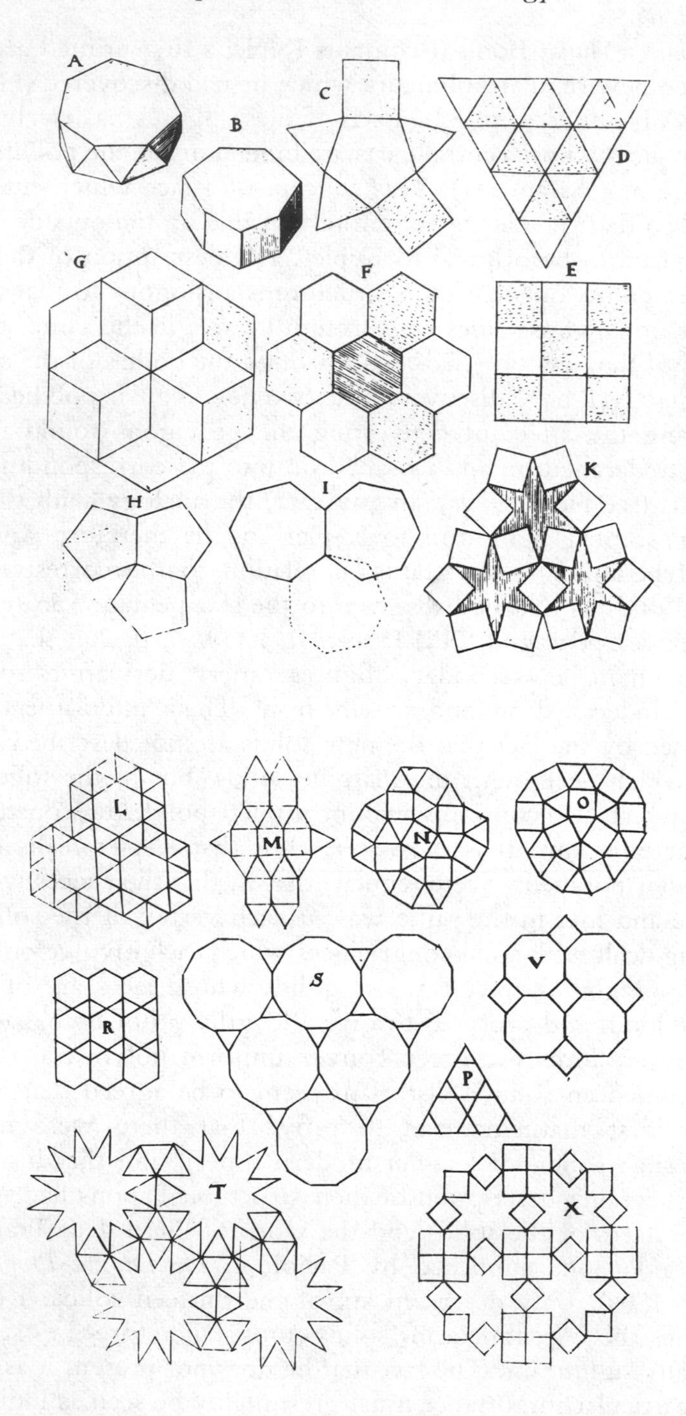

Figure 5.4 Polyhedra and tessellations (plates from HM II)

congruences were of interest mainly as showing the congruence-forming property of their constituent polygons.

In *Harmonices Mundi* Book II Kepler is mainly concerned with congruences (tessellations and polyhedra) which have uniform vertices, that is, congruences in which the polygons are arranged in the same way round every point where their corners meet. For example, in figure N (see figure 5.4) the polygons surrounding each vertex of the tessellation are, working clockwise, a square, two triangles, a square and a triangle. However, Kepler fails to distinguish this type of tessellation from the bi-uniform type we find in, say, R, where half the vertices are surrounded by two triangles and two hexagons (in that order) whereas the other half are surrounded by a triangle, a hexagon, a triangle and a hexagon. P shows a uniform tessellation which has vertices like half of those in R. This failure to distinguish uniform and bi-uniform congruences follows directly from the definitions in sections I to VI (KGW *6*, pp. 68–9). It seems particularly perverse in view of the fact that the congruences are described as being constructed by grouping polygons round a vertex, but the explanation would appear to be that, as Kepler implied in his introduction to Book II (KGW *6*, p. 67), he was really concerned with congruences only as manifestations of the varying capacity of polygons to form such structures, either in the plane or in space. The striving to enumerate all possible congruences, which makes *Harmonices Mundi* Book II an interesting piece of mathematics in the twentieth century, is apparently designed not to elucidate the properties of the congruences but rather to ensure the mathematical respectability of the classification of regular polygons in section XXIX (KGW *6*, p. 88).

Although it seems certain that the main purpose of *Harmonices Mundi* Book II must be seen in the classification of polygons described in its concluding sections, Kepler does in fact consider congruences involving polygons which do not appear in the final classification,

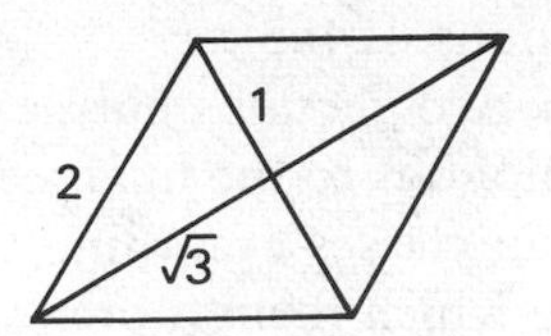

Figure 5.5 Rhombus used in tessellation G (see figure 5.4)

namely rhombi. Each of the three congruences he considers involves rhombi of a different shape. The tessellation, figure G (see figure 5.4), is made up of rhombi whose diagonals are in the ratio $1 : \sqrt{3}$, the rhombic dodecahedron, figure Vu (see figure 5.3), uses rhombi whose diagonals are in the ratio $1 : \sqrt{2}$ and the rhombic triacontahedron, figure Xx (see figure 5.3), uses rhombi whose diagonals are in the ratio $(1 + \sqrt{5}) : 2$.

As Kepler notes, the rhombus used in the tessellation is the sum of two equilateral triangles, and since he has established the existence of a tessellation made up of such triangles the existence of the tessellation of rhombi follows immediately (KGW *6*, p. 71, l. 33). Since this tessellation is merely a trivial consequence of an earlier result, it may have been included merely for mathematical reasons (or because it is a pattern sometimes seen on tiled floors?), but the case of the rhombic polyhedra is very different. Kepler does not, in fact, give a rigorous proof that they exist: the proposition in which they are described has been phrased in such a way as to make this unnecessary. Kepler's work on the rhombic solids is discussed in Appendix 4. It is fairly clear that he saw them as being derived from two of the 'primary' Platonic solids, the cube and the dodecahedron, while the cube itself, which he wished to add to the list of rhombic solids 'for its faces also have four equal sides' (HM II, sect. XXVII, KGW *6*, p. 84, l. 7), was seen as derived from the tetrahedron, the remaining 'primary' solid, by a similar construction. However, none of this is mentioned in *Harmonices Mundi* Book II section XXVII. Moreover, there is no application of these polyhedra later in the work. Nevertheless, we do know that Kepler did apply the three rhombic solids to the problem of explaining the number and spacing of the newly discovered moons of Jupiter in his *Dissertatio cum Nuncio Sidereo* (Prague, 1610, KGW *4*, p. 309, l. 304, see Chapter IV above) and in the *Epitome Astronomiae Copernicanae* (Book IV, Linz, 1620, p. 554, KGW *7*, p. 318). This suggests that Kepler intended the results of section XXVII to be as applicable as the rest of *Harmonices Mundi* Book II. Perhaps at the time he wrote Book II he intended that Book V should include a discussion of the moons of Jupiter.

Harmonices Mundi Book II ends with two sections entitled 'conclusion'. The first, section XXIX, is analogous to the final section of Book I. It places the regular polygons in classes according to their capacity for forming congruences, those which form more, or more perfect, congruences being regarded as the more noble. For example, 'the trigon and the tetragon are of the first degree because they form congruences in space as well as in the plane, both among themselves, with figures all of one kind, and also when combined with other figures' (HM II, sect. XXIX, KGW *6*, p. 88). Kepler notes, however, that a slightly different ordering is obtained if we consider only congruences in the plane (tessellations). He finally decides to divide the polygons into three classes, according to their capacity for forming congruences, the classes being named after their typical members: the octagon, the decagon and the icosigon. The section ends with an indication of the wider significance of this division: 'These classes will find their application in the choice of Aspects in Book IV' (HM II, sect. XXIX, KGW *6*, p. 89).

The second 'conclusion', section XXX, compares the classification of polygons derived in Book II, from 'congruence', with that derived in Book I, from 'demonstration'. In the Introduction to Book II Kepler had noted that since 'demonstration' related to a single polygon, considered alone, whereas 'congruence' related to combinations of polygons, which might be of more than one kind, the two properties would be expected to give rise to different classifications (HM II, Introduction, KGW *6*, p. 67, l. 4 ff). It seems likely that this preoccupation with the similarities and differences between the two classifications derives from Kepler's concern with the relationship between musical and astrological 'harmonies' (i.e. astrological Aspects) that is proposed by Ptolemy in his *Harmonica*. Whereas Ptolemy had used the musical harmonies to explain astrological ones, Kepler finds essentially independent explanations of the two sets of phenomena (see below and Field, 1984c). However, Kepler is happy to note a case in which his double scheme seems to agree with Ptolemy's unitary scheme, for he points out that 'the pentekaedecagon shows a pleasing uniformity of properties in these two respects' (HM II, sect. XXX, KGW *6*, p. 89, l. 17).

It must appear to a modern reader that in *Harmonices Mundi* Book II Kepler treats his pure mathematical discoveries as cavalierly as he had treated his laws of planetary motion in the *Astronomia Nova*. He had, however, made it clear from the beginning of *Harmonices Mundi* Book I that he was not concerned with pure mathematics but rather with applicable mathematics. The quotation from Proclus on the title page points out that mathematics is an aid to understanding the natural world, and in the Introduction to Book I Kepler had indicated that his own mathematics was conceived in exactly this spirit: '. . . I am not a Geometer working on [Natural] Philosophy, but a [Natural] Philosopher working on this part of Geometry' (HM I,

Introduction, KGW *6*, p. 20, l. 1 f). Detailed justifications for particular applications are given when Kepler comes to apply his results, in Books III, IV and V.

Music: *Harmonices Mundi* Book III

Kepler seems to have attached considerable importance to the idea of deriving musical ratios from the regular polygons which could be inscribed in a circle, thereby dividing it into a number of equal arcs. It is with a reference to this idea that he begins his introduction to *Harmonices Mundi* Book I, a passage which is, as we have seen, intended as an introduction to the whole work, only the final paragraph, with the marginal note 'the purpose of this first book', serving as a specific introduction to Book I. The first sentence of the general introduction reads 'Since today, to judge by the books that are published, there is a total neglect of the intellectual distinctions to be made among geometrical things, I thought fit to state at the outset that it is from the divisions of the circle into equal aliquot parts, by means of geometrical construction, that is, from the constructible Regular plane figures, that we should seek the causes of Harmonic proportions' (HM I, Introduction, KGW *6*, p. 15).

Kepler's interest in giving a geometrical explanation of musical ratios dates back to the time when he was writing the *Mysterium Cosmographicum*. In Chapter X of that work he attempted to provide a geometrical basis for the usual numerological account of musical ratios by showing that the 'noble' numbers involved in such ratios have a connection with the five Platonic solids: for example, by being the number of sides of the polygons that form their faces, or the number of faces of the solid (*Mysterium Cosmographicum*, Chapter X, KGW *1*, p. 37); and in Chapter XII he tried to relate the musical ratios themselves to the five Platonic solids, somewhat indirectly, through imagined divisions of the circle of the Zodiac by the polygons obtained as meridian sections of the solids, the musical ratios then being introduced as analogous to astrological Aspects,[10] both types of entity being regarded as examples of mathematical 'harmonies'. It appears that, even at the time he wrote this chapter, Kepler did not regard this account of the harmonies as entirely satisfactory, since he comments that 'because we do not know the causes of this relationship it is difficult to associate particular harmonic ratios with particular solids' (*Mysterium Cosmographicum*, Chapter XII, KGW *1*, p. 41, l. 12 ff). In the second edition of the

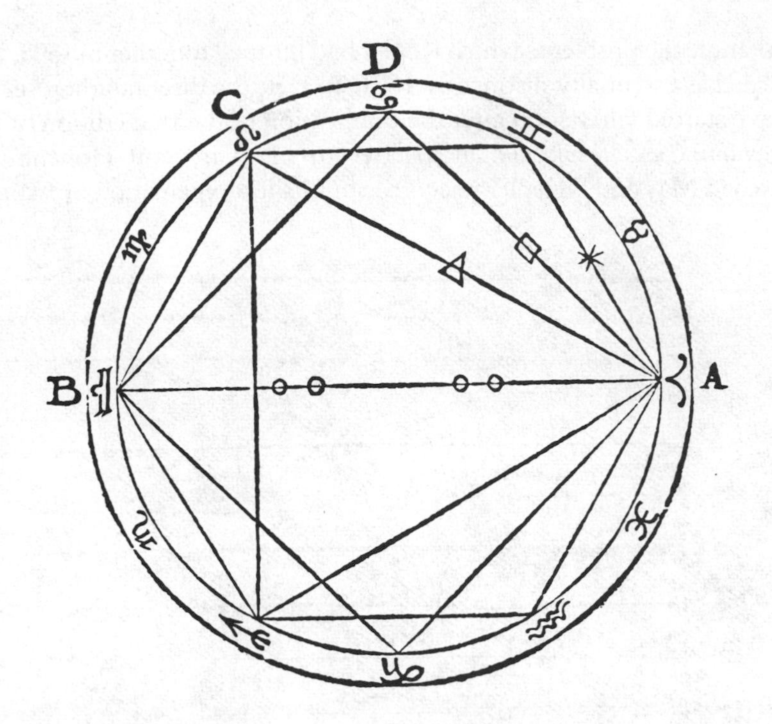

Diapaſon ſeu dupla ratio tripliciter reſpondet:	Vel	Totius circuli ad dimidium A B C ad A C, nēpe 8 ad 4. A C B ad A D, ſex ad tria.
Diapente ſeu ſeſquialtera, item tripliciter:	Vel	Totius circuli, ſeu 12.8 D A B, ideſt, 9 ad 6 A B, ideſt, 6 ad A C, ideſt 4.
Diateſſaron ſeu ſeſquitertia, item tripliciter:	Vel	Totius circuli ad A B C D, ſeu 12 ad 4 in A B D. A B C ad A B, ideſt 8 ad 6. A C ad A D, 4 ad 3.

Diapaſon & Diapente, item tripliciter, Biſdiapaſon uero dupliciter Tonus ſemel.

Figure 5.6 Ptolemaic Aspects and Consonances, from Ptolemy, *Harmonica*, trans. Gogava, Venice, 1562, page 144.

Mysterium Cosmographicum (Frankfurt, 1621) this chapter has acquired thirty-nine notes, whose total length is slightly greater than that of the original chapter. These notes are mainly concerned to disentangle the problems which Kepler had lumped together in 1595, but regarded as essentially distinct by 1621. In fact, the disentangling seems to have started fairly soon after the publication of the first edition of the *Mysterium Cosmographicum*: in a letter to Herwart von Hohenburg, written in May 1599, Kepler rejects the published explanation of Aspects

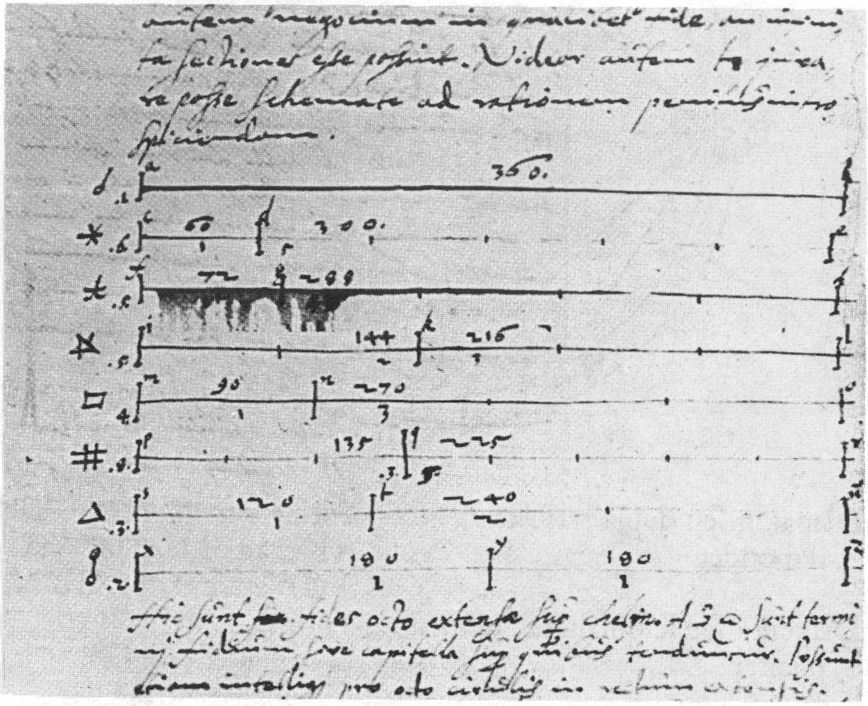

Figure 5.7

Correspondence of astrological Aspects and harmonic ratios, from a letter written by Kepler to Herwart von Hohenburg (30 May 1599).

Ptolemaic Aspects:

AB, conjunction (0°, undivided string), used as an Aspect by Ptolemy, though he does not define it as such;
CDE, sextile (60°, 1:5);
MNO, quadrature (90°, 1:3);
STU, trine (120°, 1:2);
XYZ, opposition (180°, 1:1).

New Aspects:

FGH, quintile (72°, 1:4);
IKL, biquintile (144°, 2:3);
PQR, sesquiquadrature (135°, 3:5).

and suggests an alternative which does not refer to the Platonic solids, but instead derives Aspects from musical ratios among the arcs into which the circle of the Zodiac is divided by bodies that are at Aspect to one another. This procedure is clearly derived from Ptolemy's *Harmonica*, of which Kepler seems to have had some knowledge at least since 1595 (there is a passing reference to the work in *Mysterium Cosmographicum* Chapter XII) although he had not yet managed to obtain a copy of the book (see Klein 1971). Ptolemy's scheme for relating consonances and Aspects is shown in figure 5.6. In his letter to Herwart, Kepler shows a rather similar scheme, but with the Zodiac opened out to resemble the string of the traditional monochord (see figure 5.7).[11] (It will be noted that in order to get the correct musical ratios Kepler has had to admit three new Aspects.) In this letter Kepler seems to regard the musical ratios as merely given, but a later letter to Herwart, written in August 1599, contains a fairly substantial attempt to explain their origin. The letter is so long that Kepler has divided it into numbered sections. The fifth begins

> 'On the causes of musical harmonies. And since you are interested in my discussion of Aspects I shall add something here on that too. I changed my shape a thousand times, like Proteus, first to show by Geometry that there were just seven Musical harmonies, and then, on your account, to separate the cause of Aspects from the cause of the harmonies, leaving Astronomy no more Aspects than are usually accepted. I have hold of both parts, but as one has a wolf by the ears . . .'[12]

Kepler lists the musical ratios he is looking for, dismisses the idea of seeking an explanation in arithmetic ('For in this matter nothing can come of Arithmetic, since whatever fitness numbers have arises from Geometry and the things that are numbered'), dismisses lines as being infinite (though it is not clear in what sense: his words are '*In lineis nulla finitio est*'), and then looks to the circle: 'Among surfaces the circle establishes a kind of infinite finiteness, and there is regularity [i.e. symmetry?] in what may be inscribed in a circle' (letter 130, l. 343, KGW *14*, p. 30). This passage does not really amount to an explanation or even an argument in favour of certain ideas, it is no more than a series of assertions addressed to a like-minded reader. The same is true of the paragraph that follows, in which Kepler states which polygons he will need to exclude, and gives grounds for excluding them, such as that, in the case of the decagon and the pentagon, the square of the side is not commensurable with the square of the diameter of the circle (letter 130, l. 352 ff, KGW *14*, p. 30). However, he found he could not construct a satisfactory scheme by this method, so he looked at the sizes of the angles of the regular polygons concerned, thinking that he might be able to exclude all the figures in which three vertical angles came to more than four right angles 'though it is not clear why Nature should take special account of this property in setting up ratios' (letter 130, ll. 397–402, KGW *14*, p. 31). This idea naturally led to tessellations, since if polygons are to form a tessellation the angles which meet at each of the vertices must add up to exactly four right angles (letter 130, l. 419 ff, KGW *14*, p. 32). The work on tessellations, which was eventually to find its application in connection with astrological Aspects in *Harmonices Mundi* Book IV (see below), thus makes its first appearance as a possible explanation for musical ratios. In fact, the next section of this letter turns to Aspects, the problem being, as Kepler sees it, to explain why, when it comes to Aspects, astronomy admits fewer ratios (between arcs of the Zodiac) than are admitted (between lengths of strings) in music (letter 130, ll. 573–6, KGW *14*, p. 37). As in the previous sections of the letter, all the suggested explanations are derived from Geometry. In two other letters written in the summer of 1599, Kepler gave similar, but rather more cursory, accounts of the relation between Aspects and musical ratios to Edmund Bruce (18 July 1599, letter 128, KGW *14*, p. 7) and to Maestlin (19 Aug. 1599, letter 132, KGW *14*, p. 43).

Between them, these two letters to Herwart contain all the elements which were to be reassembled to form the geometrical explanations of musical consonances and astrological Aspects nearly twenty years later in *Harmonices Mundi* Books III and IV, where musical ratios are derived from those polygons whose sides are most closely related to the diameter of the circle in which they are inscribed, while the astrological ratios are derived from those polygons which will fit together to form tessellations or polyhedra. Nevertheless, it is clear that in 1599 Kepler's purpose is still primarily to explain the musical ratios, which are then to be used to explain the astrological ones. This is the pattern set by Ptolemy in his *Harmonica*. It is not clear exactly when Kepler was to break away from it and seek instead to give essentially separate explanations of the musical and astrological ratios such as are found in the *Harmonice Mundi* (see below and Field, 1984c).

The traditional explanation of the origin of musical ratios appealed to the alleged special nature of the first few integers; how many integers were involved depended upon the system of ratios that had to be explained: 'Pythagorean' intonation or 'just' intonation.

Pythagorean intonation was derived from ratios between the integers 1 to 4. To obtain an interval of a fourth, the ratio of the lengths of the strings had to be 3 : 4, for a fifth 2 : 3 and for an octave 1 : 2. Fitting in equal tones between the first note and these three fixed points gave dissonant thirds and sixths, corresponding to the ratios 64 : 81 and 16 : 27. This Pythagorean system was under serious attack by the second half of the sixteenth century, since consonant thirds and sixths were a necessity for the polyphonic music which was being written by composers such as Orlando di Lasso (1531–94). However, the old system found a vigorous defender in Vincenzo Galilei, whose *Dialogo della musica antica et della moderna* (Florence, 1581) Kepler apparently read with enjoyment, for its detailed account of ancient theories and for the author's polemical skill, while disagreeing with its fundamental thesis.[13]

The system of 'just' intonation was codified by Gioseffo Zarlino in his *Istitutioni Harmoniche* (Venice, 1558). It accepts thirds and sixths as consonances, corresponding to ratios of string lengths of 4 : 5 and 5 : 6 for the major and minor thirds, and 3 : 4 and 5 : 8 for the major and minor sixths, while the fourth, fifth and octave are the same as in the Pythagorean system. Kepler accepted the 'just' system, apparently on the grounds that observation showed thirds and sixths to be consonant (see Walker 1967, 1978). Zarlino's system is substantially identical with that described by Ptolemy in his *Harmonica* and Kepler usually refers to it as being Ptolemy's system, with the result that Zarlino's name only occurs once in *Harmonices Mundi* Book III.

Zarlino's system represented the musical orthodoxy of its day, but by the early years of the seventeenth century it was no longer regarded as adequate by all practising musicians (see Palisca 1961). In response to an attack upon him for his use of dissonances, Claudio Monteverdi (1567–1643) appended a few lines of Italian prose to the fourteen pages of music for the *basso continuo* parts to his *Fifth Book of Madrigals for Five Voices*:

> '. . . I have written a reply to make it clear that I do not compose my works at random, and as soon as it is rewritten it will be published under the title Second System (*Seconda Practica*), or Perfection of Modern Music, at which perhaps some may be surprised, not believing that there is any other system than that described by Zarlino; but be assured that in connection with consonances and dissonances there is another explanation

(*consideratione*), different from that already given, which, in accordance with reason and the evidence of the senses, defends the modern style of composition . . .' (C. Monteverdi *Basso continuo del quinto libro de le madrigali a cinque*, Venice, 1605, p. 21).

In fact, Monteverdi's promised work never appeared in print.

Since Kepler accepted Zarlino's system, it comes as no surprise that his occasional references to musical compositions include three to works by Orlando di Lasso and none to works by Monteverdi. In the matter of music theory, Kepler was on the side of orthodoxy rather than standing up to be counted as a partisan of the *avant garde*.

Harmonices Mundi Book III begins in the Aristotelian manner, with a historical introduction which defines the problem and then proceeds to demolish the theories put forward by the writer's predecessors. It is, of course, 'the Pythagoreans' who are ostensibly in the firing line. However, their theories are taken to include the numerical series 1, 2, 4, 8 and 1, 3, 9, 27 used to describe the structure of the strips of the Same and Different which form the circles corresponding to the celestial equator and the ecliptic in the model of the heavens described in *Timaeus* (34b–36d), so Plato is clearly included in the company, though his name is not actually mentioned.

There follows a discussion of the significance which the Pythagoreans attached to the 'Tetractys', i.e. the integers 1 to 4. Kepler includes (in Latin translation) a long quotation from Camerarius' commentary on the *Carmina Aurea* (Camerarius *Libellus Scholasticus*, Basel, 1551, pp. 205–8; though its title is Latin, the work itself is written in Greek). The margins of the quotation are strewn with triangular, square and oblong arrays whose principal element is the figure 1. Having quoted Camerarius' opinion of 'the ancients' Kepler next turns to the number philosophy of Hermes Trismegistus, '(whoever he may have been)', as expounded in *Pimander*, which he judges to be very close to the opinions of the Pythagoreans. Adapting Aristotle's comment on a passage in Plato's *Republic*, Kepler notes 'there is no doubt that either Pythagoras is Hermetising or Hermes is Pythagorising' (KGW *6*, p. 99, ll. 1–2).

In *Giordano Bruno and the Hermetic Tradition* Yates quotes the passage concerning Hermes, *in extenso* and in Latin, commenting that it does not allow one to decide whether or not Kepler knew of Isaac Casaubon's re-dating of Hermes (Yates, 1964, p. 442 – Casaubon's work was published in London in 1614). However, the fact that Kepler (1) discusses Hermes after 'the Pythagoreans', (2) draws attention to the parallels with the Old Testament ('Moses') and St John's Gospel (parallels which were the subject of comment by Casaubon, see Yates, 1964) and (3) explicitly mentions that Hermes may be 'pythagorising', all these suggest that Kepler may at least have heard that doubt had been cast upon the very ancient date usually ascribed to the Hermetic Corpus. They also suggest that he did not consider such doubts entirely unreasonable. This one page concerning *Pimander* appears to be the only reference to Hermes in the *Harmonice Mundi*, except in connection with Kepler's repeated assertions that his own mathematics is quite unlike the Hermetic mathematics of Fludd. Since Yates regards the summary of *Pimander* as evidence that Kepler had studied the work with care (Yates 1964, p. 444), we may presumably accept it as a fair account of Hermes' opinions, though it is given only as a prelude to pointing out that Kepler disagrees with them.

Returning to the particular problem of musical ratios, Kepler first criticises the Pythagoreans for relying solely on the properties of numbers to decide the matter: 'The Pythagoreans were so committed to this philosophy through Numbers that they paid no attention to the judgement of their ears, though their ears had provided the original basis for this study. Instead they defined what was harmonious and inharmonious, what was consonant and dissonant, solely by means of their Numbers, doing violence to the natural faculty of hearing' (KGW *6*, p. 99, ll. 12– 7). Ptolemy's modification of the Pythagorean system admitted ratios involving integers greater than four, but did not consider that major and minor thirds and sixths should be regarded as consonances, '[intervals] which all modern musicians who have a good ear assert to be consonances' (HM III, p. 8, KGW *6*, p. 99, l. 35). Kepler's first objection against the traditional systems is thus an empirical one: that they are not in agreement with observation. (As Walker (1967, 1978) has noted, Kepler's musical work repeatedly makes such appeals to the results of observation and lays great stress upon their importance in providing a proper basis for musical theory.) However, the next objection is philosophical: that the systems are deduced from the properties of pure numbers.

> 'Further, if they were completely equal in scope, the cause taken from abstract Numbers and the effect of Consonance, it would not be absurd to see this cause as archetypal, evidence that the Father of creation, Eternal mind, contemplating the Numbers,

> had taken from them the Idea of tones and intervals . . . However, it is not yet sufficiently clear why the numbers 1, 2, 3, 4, 5, 6 etc. give rise to Musical intervals, while the numbers 7, 11, 13 and the like do not give rise to them. Nor do the numbers, as numbers, show by themselves why this should be so' (HM III, p. 8, KGW *6*, p. 100, ll. 3–11).

Kepler's objection is thus not against the use of numbers as an archetypal cause but against the inadequacy of the particular numerological theory that has been advanced to account for musical consonances. It even appears from his next sentence that he might have been prepared to accept something like the two series of numbers put forward in *Timaeus*, if such a theory had been able to account for the observed consonances: 'For the cause deduced from the first three numbers, and the family of squares and cubes derived from them, is not a cause (*causa est nulla*); since it excludes the number five, which, arising as it does among the musical intervals, cannot be deprived of its right of citizenship' (KGW *6*, p. 100, ll. 11–13). He has, however, a further, more radical, objection to numerological theories of consonance: 'But it is not enough for the theoretician to know that the Numbers 1, 2 and 3 are symbols of the Principles which make up Natural things. For an interval is not a natural thing, but a Geometrical one; so unless these numbers were to number something else, more closely related to intervals, a philosopher could not put his faith in this cause, but would suspect it of not being a cause' (KGW *6*, p. 100, ll. 15–20).[14] Experiments carried out later in the seventeenth century were to show that there was, in fact, a physical basis for associating small integers with consonances: the integers represented multiples of wavelengths in the vibration of a string (Palisca 1961) – and it so happens that the human ear hears a combination of notes as pleasant, 'consonant', if the ratios of the wavelengths involved are expressible in terms of small numbers, such as 1 : 2 and as unpleasant, 'dissonant', if the ratios involve larger numbers, such as 17 : 23. This fact is believed to be connected with the length of time between 'beats' in the compounded wave. In 1618, however, Kepler's objection was still an entirely valid one, as it had been twenty years previously, when, according to his own account, he first began to think about the problem (HM III, p. 8, KGW *6*, p. 100, l. 21 ff). It is curious that Kepler should date his interest in this aspect of musical theory only to 1598, after the publication of the *Mysterium Cosmographicum*, since, as we have seen, Chapter XII of that work began by tracing a connection between meridian sections of the regular polyhedra and Aspects, then used musical consonances by way of analogy and attempted to relate the generalised 'harmonies' back to the polyhedra. Presumably Kepler regarded this work as a stage in the development of his explanation of Aspects, which was the main subject of the chapter, rather than the beginnings of a geometrical explanation of musical harmonies.

The justification for turning to geometry to explain consonances and musical ratios is stated quite briefly: 'For since the terms of the ratios which correspond to consonant intervals are continuous quantities, the causes which separate consonances from dissonances must also be sought among the family of continuous quantities, not among abstract Numbers, since they are discrete quantities' (HM III, p. 9, KGW *6*, p. 100, ll. 30–3). (Kepler always uses *numerus*, the word translated here as 'number', in the sense in which he is clearly using it here, namely as a translation of Euclid's ἀριθμος, to mean an integer. Like Euclid, he uses geometry to handle continuous quantities, though, as we have seen, he was prepared to try the new technique of algebra when the problem of inscribing a regular heptagon in a circle proved to be insoluble by means of geometry (HM I, sect. XLV, KGW *6*, p. 50).)

Having justified his use of geometry rather than arithmetic, Kepler turns to the reason why geometrical knowledge, as described in *Harmonices Mundi* Book I, should decide which ratios give rise to consonances:

> 'And since it was Mind which so formed human Souls that they should take pleasure in some particular interval (which is the true definition of consonance and dissonance), both the differences between one interval and another and the causes which make these intervals harmonic, must be mental and intellectual in nature, their nature doubtless being such that the terms which make consonant intervals are knowable by proper demonstration, and the terms of dissonances either knowable, but not by proper demonstration, or not knowable. For if they are knowable they can thus enter into the Mind and take part in constructing the archetype, but if, however, they are unknowable (in the sense that was explained in Book I) they will thus remain outside the Mind of the eternal Artificer and in no way contribute to the Archetype' (HM III, p. 9, KGW *6*, p. 100, l. 33 to p. 101, l. 3).

Whether a figure is knowable, that is, whether it can be inscribed in a circle by means of a straight edge and compasses, is thus taken to be a fundamental property of the figure, determining whether it can contribute to the archetype according to which the World is created. We have seen, in our discussion of *Harmonices Mundi* Book I, that Kepler regarded the property of being knowable as a criterion of nobility, indicating the closeness of a figure's relation to the circle, and thus its fitness to contribute to the archetype, but when he restates the condition in the form just quoted it is rather difficult to convince oneself that he is not putting arbitrary limits to God's powers by restricting Him to using only a straight edge and compasses. Presumably this restriction seemed less arbitrary when there really was no accurate method of constructing a regular heptagon, and when algebra had not yet advanced to the point where obtaining an equation whose solution is the side of the polygon seems at least as valid a way of knowing the answer as does constructing the polygon in a circle. In any case, the great gulf between knowable and unknowable polygons seems to have appeared real enough to Kepler. He therefore sets about applying the hierarchy of polygons, obtained in Book I, to the task of characterising ratios of integers as corresponding either to consonances or to dissonances, taking the relations of knowability to consonance as axioms. For example,

> 'Axiom I
>
> The diameter of the circle, and the sides of the basic figures described in Book I which have a proper demonstration, determine a part of the circle which is consonant with the whole circle'(HM III, Ch. I, KGW *6*, p. 102); and
>
> 'Axiom II
>
> By whatever degree the demonstration of the side differs from the first degree, the part of the circle cut off by the side will, in consonance with the whole circle, differ from the most perfect consonance, unison, by the same degree. Or, as is the place of the figure, whose side is in question, among the other figures, so is the place of its consonance among the other consonances' (KGW *6*, p. 103).

Axioms III and IV are also of this type, but the final three axioms, numbers V, VI and VII, are concerned with consonances as such rather than with regular polygons. After the axioms, there is a passage of philosophical comment, rather like the passage which followed the mathematical description of the Platonic solids in *Harmonices Mundi* Book II, section XXV (KGW *6*, p. 80). It begins

'To contemplate these Axioms, particularly the first five, is sublime and Platonic and resembles Christian Faith concerning Metaphysics and the doctrine of the Soul. For Geometry, whose relevant parts have been dealt with in the first two Books, Geometry, coeternal with God and shining in the divine Mind, gave God the pattern, as explained in the introductory section of this Book, by which He laid out the World so that it might be Best and Most Beautiful, and finally most like the Creator' (HM III, Ch. I, KGW *6*, p. 104, l. 35 to p. 105, l. 3).

The references, nestling somewhat incongruously among the rhetorical phrases, serve as a reminder that this statement of faith is, in fact, the preamble to a philosophical discussion of the place of harmonies in the scheme of things, after which Kepler turns to the main subject of *Harmonices Mundi* Book III Chapters I and II, namely the deduction of musical ratios.

This deduction first takes the form of a series of propositions relating the ratios of different arcs of a circle one to another, and Chapter I ends with a table showing consonances and dissonances between the whole circle and parts of the circle cut off by the inscription of a diameter or a regular polygon with 3, 4, 5, 6, 8, 10, 12, 16, 20 or 24 sides. Kepler considers both the ratio between the arc subtended by a given number of sides and the whole circle, and the ratio between the remaining arc and the whole circle (HM III, Ch. I, KGW *6*, p. 113).

Musical notation makes its first appearance in Chapter II, and from that point the propositions take an explicitly musical form. For example, the third proposition of the chapter reads 'Proposition XI Dividing a string into two parts, in triple proportion to one another, gives a harmonic ratio' (KGW *6*, p. 115; the numbering of propositions is consecutive throughout Book III). The accompanying diagrams show the division of straight lines, not circles, but the proofs appeal to the propositions which were summarised in the table at the end of Chapter I and to the propositions concerning regular polygons inscribed in a circle, which were proved in *Harmonices Mundi* Book I. The final proposition of Chapter II, Proposition XIX, is that 'After that defined by the octagon there are no more Harmonic divisions of a string' (HM III, Ch. II, Prop. XIX, KGW *6*, p. 118). From this proposition, proved by appealing to the indemonstrability of figures with more than eight sides, and ruling out the pentekaedecagon because it has only an improper demonstration (see HM I, sect. XLIV, KGW *6*, p. 46), Kepler proceeds to draw the corollary that 'there are seven harmonic divisions of a string, not more' (KGW *6*, p. 118, l. 30), these seven corresponding to the ratios required by 'just' intonation, the Ptolemy–Zarlino system, including 4 : 5 and 5 : 6 for the major and minor thirds and 3 : 5 and 5 : 8 for the major and minor sixths.

Kepler has no difficulty in explaining the ratios corresponding to the old Pythagorean consonances, namely the ratios 1 : 2, 2 : 3, 3 : 4. These clearly correspond to divisions of the circle by the diameter (for 1 : 2), the equilateral triangle (for 2 : 3) and the square (for 3 : 4), figures which are demonstrable and have sides which are commensurable, or commensurable in square, with the diameter of the circle.

The four extra consonances required by 'just' intonation present slightly more of a problem, since they involve a ratio that can only be produced by means of the pentagon, namely 4 : 5 (Kepler cannot use an enneagon for this ratio because the enneagon is not demonstrable). Kepler argues for the consonance-forming power of the pentagon on the grounds that the incommensurability between its side and the diameter of the circle involves the Golden Section, the Divine Proportion. He makes this point briefly at the end of *Harmonices Mundi* Book I (KGW *6*, p. 63, l. 36 to p. 64, l. 2) and argues it in more detail, connecting the Divine proportion with the principle of generation, in Book III, Ch. XV.[15] The pentagon gives the ratios 4 : 5 and 3 : 5 (the latter can also be obtained from the octagon). The ratios corresponding to the other two 'just' consonances, 5 : 6 and 5 : 8, present little difficulty. They can be derived from the hexagon and the octagon, figures which are closely related to the equilateral triangle and the square (since their sides can be obtained by bisecting the arcs cut off by the sides of the latter figures) and therefore share their power of determining consonances.

Having achieved this geometrical explanation of the consonances, Kepler rather disconcertingly goes on to present them in a form which emphasises arithmetical relations between them. He shows that the ratios can be obtained by the following process: Starting with $\frac{1}{1}$, one adds numerator and denominator to obtain a new denominator, namely 2, and then uses the old numerator and denominator as two new numerators, giving $\frac{1}{2}$ and $\frac{1}{2}$. Operating on $\frac{1}{2}$ in the same way gives $\frac{1}{3}$ and $\frac{2}{3}$, and these fractions in their turn generate $\frac{1}{4}$, $\frac{3}{4}$ and $\frac{2}{5}$, $\frac{3}{5}$. The process can be continued until one obtains a denominator which is the number of sides of an indemonstrable polygon. A little table of these 'harmonic' fractions and an explanation of this arithmetical method of obtaining them forms the second part of the corollary to Proposition XIX.

Armed with his consonances, Kepler would appear to be free to move on to music theory proper, but he first includes a chapter on what was (and is) usually meant by 'harmonic' proportion, namely that if the number b is the harmonic mean of a and c then

$$\frac{1}{b} = \frac{1}{2}\left(\frac{1}{a} + \frac{1}{c}\right).$$

(Kepler, of course, does not use the algebraic formulation.) He has no difficulty in showing that this method of division will lead to dissonant ratios, and he accordingly rejects it. Unfortunately, as Walker notes, this form of harmonic division is a direct indicator of the physical cause of consonance (Walker, 1978, p. 50). The reason for Kepler's rejection of what is now seen as a revealing formulation lies in his philosophical assumptions: that consonances, being beautiful, were the significant elements in music, revealing the beautiful Archetype according to which God had created both the World and the correspondingly sensitive human soul.

About a third of the way through this chapter on the harmonic mean, after the first paragraph of page 29 in the original edition (KGW *6*, p. 121, l. 11), the printer inadvertently left out some pages, according to the account of the matter at the top of page 86 (KGW *6*, p. 186, l. 1). The material omitted is entitled 'A political digression on the three types of mean'. Since this digression is about seventeen pages long, the printer may well have felt he was doing Kepler's readers a good turn by transferring it to the end of Book III rather than putting it into a chapter whose length was otherwise about three pages.

The 'Political Digression' is based on the final chapter of a work by Jean Bodin, first published in French in 1576. Kepler seems to have used the Latin edition of 1586, entitled *De Republica Libri VI* (see Caspar's note, KGW *6*, p. 538). Bodin's title seems to echo Plato, but the work is in fact very different from Plato's *Republic*, being concerned with statecraft rather than the construction of a system which will illuminate the concept of justice. Bodin comes to this concept only in his very last chapter (*La République*, Paris, 1576, Book VI, Ch. VI, pp. 727–59) which begins: 'It remains to end this work by considering Justice, the main basis of any Republic, and of such importance that Plato even called his ten books on the Republic a treatise on Justice, though he wrote of it rather as a Philosopher than as a legislator or a lawyer.' Although the title of Bodin's final chapter proclaims the author's intention of making use of arithmetic, geometric and harmonic means his use of mathematics is somewhat tentative, taking the form of a few numbers introduced here and there by way of illustration. Moreover, the mathematical basis of these illustrations is not a sound one: Bodin gives a vague definition of what is meant by a harmonic mean, but vague as it is, it will not cover the correct definition. Kepler quotes it with the comment 'Now this definition is not correct' (KGW *6*, p. 187, l. 27) and then proceeds to explain why not, at length. Later, he discusses 24 sections of Bodin's work in which the reasoning is vitiated by the use of this incorrect 'harmonic' mean, making his own suggestions as to how they should be modified (KGW *6*, pp. 189–205). It is clear that Kepler is thoroughly in favour of Bodin's approach. His professional disgust is directed only at the incorrect definition, not at the importation of the mathematical method of *Timaeus* into the territory of the *Republic.*[16]

The remainder of *Harmonices Mundi* Book III, Chapters IV to XVI, is concerned with music theory proper. Kepler deals with such technical matters as the types of scale (Ch. IV) and the problem of tempering (Ch. VIII). In the course of the latter discussion he suggests a new way of tuning the strings of a lute, which, he claims, is better than the tuning proposed by Vincenzo Galilei (HM III, Ch. VIII, KGW *6*, p. 143). Caspar has shown that, over the years that had elapsed since Kepler had sketched out his first plan for a work on Universal Harmony, he had studied the writings of the musical theorists of his own day as well as the works of Euclid, Ptolemy, Boëthius, Jordanus Nemorarius and Jacques Lefèvre of Étaples (Caspar, 1940, p. 477). It is therefore clear that Kepler intended his own work to be a serious contribution to musical theory, and historians have generally accepted it as such.[17] For our present purpose, however, the interest of Kepler's theory lies in the nature of its mathematical basis, and in its application to the problem of explaining the observed structure of the Solar system, which will be considered in a later section of this chapter.

Astrology: *Harmonices Mundi* Book IV

As we have seen, astrology does not play a very important part in the *Mysterium Cosmographicum*, and Kepler's later notes suggest that by 1621 he regarded two of the three brief astrological chapters as irrelevant to the main purpose of the book. In contrast, the astrology of *Harmonices Mundi* Book IV is not only much more mathematically sophisticated than that in the earlier work, it is also clearly an integral part of Kepler's description of his scheme of universal harmonies, of which a slightly modified version of the theory described in the *Mysterium Cosmographicum* is also a part. The astrology in *Harmonices Mundi* Book IV has much the same status as the astronomy in Book V: the mathematical results obtained in the earlier part of the work are used to explain the existence of various Aspects in the same way as they are used to explain, for example, the eccentricities of planetary orbits. However, Kepler limits his astrological explanation to giving an a priori account of the number and relative power of Aspects, thus ignoring much of the traditional subject-matter of astrology. We know, of course, that Kepler rejected many of the standard astrological beliefs of his day, such as the belief that particular characters should be ascribed to individual signs of the Zodiac.[18] Nevertheless, it appears that when he was working on *Harmonices Mundi* Book IV Kepler considered it reasonable to ascribe a particular character to each planet, in terms of the colour of the light it sends to the Earth (HM IV, Ch. VII, KGW *6*, p. 279, l. 30 ff); and in 1621, when he wrote the notes for the second edition of the *Mysterium Cosmographicum*, he described Chapter IX, which accounts for the astrological characters of the planets by reference to the corresponding regular polyhedra, merely as 'a digression' (KGW *8*, p. 59), rather than dismissing it entirely, as he had dismissed other parts of his work (see Chapter IV above).

It seems that Kepler had taken a sceptical attitude to traditional astrology from quite early years. In March 1598 he wrote to Maestlin '. . . I am a Lutheran astrologer, I throw away the nonsense and keep the hard kernel' (Kepler to Maestlin, 15 March 1598, letter 89, l. 177, KGW *13*, p. 184). However, it appears that what seemed 'nonsense' and what 'hard kernel' did change over the years, so that, as Simon (1975 and 1979) has shown, the *Confessio Augustana* of Kepler's alleged astrological Lutheranism was more a collection of principles – principles closely akin to those governing his astronomical beliefs – rather than a definite body of dogma. One of these principles was to demand that a theory should account for the observations, with an appropriate degree of precision: to 5 minutes of arc for the position of Mars, rather roughly when it came to astrological predictions of the weather, since the weather was clearly dependent not only on the Aspects operating from the heavens but also on such things as the local disposition of the Earth. As Kepler explained:

> '. . . the greatest variation [sc. in response to Aspects] is in the disposition of the bodies acted upon, particularly the disposition of the Earth, which is different in different parts and at different times. For when, in Spring, humours are abundant in the Northern hemisphere, because the Sun is moving higher [i.e. the days are lengthening], as I explained above, then even the least powerful Aspect, of any planets, will excite such humours into activity and make them produce a quantity of vapour that will fall as showers' (*De Fundamentis Astrologiae Certioribus*, Prague, 1602, Thesis XLIV, KGW *4*, p. 24, ll. 32–8).

Furthermore, Kepler believed that the theory that the weather was affected by Aspects was amply confirmed by observation. Indeed, he made many observations of the weather to check the theory. Some of these observations were published in *Tertius Interveniens* (Frankfurt, 1610), where Kepler used them to prove, against Feselius, that a conjunction of Saturn and the Sun causes coldness in the weather:

> 'In the year 1592, on 9 July, New Style, [such a conjunction took place] in Cancer, when I had not yet begun to take note. But Chytraeus writes that the whole Summer, particularly at that special time, was cold and wintry. In 1593, 24 July, in the first part of Leo. There was a great confusion of Aspects. For the Sun, Venus and Saturn were in conjunction, Mars was at sextile to Jupiter and, further, Mercury was moving back from being at trine to Mars. On 20, 21 and 22 July there was much rain and hail. The 23rd was cloudy . . .' (*Tertius Interveniens*, Frankfurt, 1610, CXXXIV, KGW *4*, p. 254.)

Kepler cites seventeen such examples, giving varying quantities of detail, the last one dating from 22 and 23 January 1609. Eleven of the examples refer to the months from October to January, when one would naturally expect the weather to be cold, or, in Kepler's terms, when the disposition of the Earth made it susceptible to influences that would cause an increase in cold. Kepler's success in obtaining observational confirmation of his belief in the efficacy of Aspects may be partly due to the subjectivity of the data, but another explanation also presents itself: Aspects are so numerous that for any given change one could almost certainly find an appropriate recent Aspect. This objection in fact occurred to one of Kepler's regular correspondents, the physician Johann Georg Brengger, who mentioned it in a letter to Kepler dated 7 March 1608 (KGW *16*, letter 480).

Kepler himself believed that the influence of Aspects upon the weather was so strongly marked that one could hope to decide by observation whether a proposed new Aspect should be accepted or not. We have seen that in a letter to Herwart in 1599 he apparently felt uneasy because his theory of the origin of Consonances seemed to require the acceptance of some unconventional Aspects (Kepler to Herwart, 30 May 1599, letter 123, ll. 359–416, KGW *13*, pp. 349–50 see figure 5.7 above), but by the end of 1601 he was willing to propose in print the acceptance of these three new Aspects (corresponding to separations of one fifth, two fifths and three eighths of the Zodiac) and appealed to observation in support of his proposal (*De Fundamentis Astrologiae Certioribus*, Prague, 1602, Thesis XXXVIII, KGW *4*, p. 22). Similar observational justification of unconventional Aspects is suggested in a letter to Brengger, written in 1607 (Kepler to Brengger, 30 Nov 1607, letter 463, l. 36 ff, KGW *16*, p. 85). Brengger's sceptical reply was answered with an account of a striking example of unseasonal weather apparently caused by a quintile aspect between Saturn and Jupiter.[19] However, the problem of the number of Aspects compared with the number of Consonances continued to worry Kepler (as he mentions in *Tertius Interveniens*, see below). The fact that he continued to find a discrepancy between Consonances and Aspects may seem surprising in view of the ease with which he seems to have convinced himself of the efficacy of certain new Aspects, such as the one where planets were separated by one fifth of the Zodiac. One might have expected that observational confirmation of any putative Aspect would have been easy to come by. Presumably the explanation for Kepler's failure to find observational confirmations must be that, like many a modern astronomer or cosmologist, he was more easily persuadable on some points than on others.

It appears that by 1610 he had already rejected the straightforward (Ptolemaic) analogy between Aspects and Consonances which he had set out in his letter to Herwart in 1599,[20] for in *Tertius Inter-*

veniens he remarks 'So I finally acknowledged the difference between Music and Astrology, and then I wondered why, indeed, I should not particularly notice the sesquiquadrate [i.e. $\frac{3}{8}$], decile [$\frac{1}{10}$] and tridecile [$\frac{3}{10}$] Aspects, and should take such emphatic notice of the semisextile [$\frac{1}{12}$], and on the other hand why the octagon and decagon and the chord subtending three tenths should be so noble, and radically (*schier*) nobler than the dodecagon' (*Tertius Interveniens*, Frankfurt, 1610, LIX, KGW *4*, p. 205, ll. 16–21). The formulation of the problem indicates the direction in which Kepler's thoughts were turning in search of a solution, and his next sentence duly begins 'So then I started to hunt through Geometry . . .'.

By the time Kepler came to write *Harmonices Mundi* Book IV his list of Aspects had been subject to minor adjustments, but it still included ratios which were not to be found among musical consonances, and it therefore required an explanation in terms of a geometrical hierarchy slightly different from that used to explain the consonances. However, Kepler is careful to make it clear that Aspects, like Consonances, must be seen as 'harmonies', as they were by Ptolemy, and the first three chapters of Book IV are accordingly taken up with the problems of the essence of harmonies (in sensible and abstract entities), the soul's faculty for perceiving harmonies, and in what things harmonies may be perceived (by God or by Man). The fourth chapter deals with the distinctions to be made between the musical harmonies considered in the last book and the astrological ones to be considered in the present one. These distinctions are all physical, in the wide sense in which Kepler uses the word (see above p. 53), and are summed up succinctly in the marginal notes: 'The harmonies in this book are narrower' (i.e. they do not interact as musical harmonies do), 'These harmonies concern angles', 'In the form of arcs of the Zodiac', 'They are not truly celestial', 'but terrestrial' (i.e. perceived from the Earth, being angles made at the Earth), and so on (HM IV, Ch. IV, KGW *6*, pp. 234–5). Harmonies are perceived by the human soul, and since the Earth also appears to respond to harmonies it seems to Kepler that the Earth too must have a soul (HM IV, Ch. IV, KGW *6*, pp. 236–7). He had made this same suggestion, on the same grounds, many years earlier, in *De Fundamentis Astrologiae Certioribus* (Prague, 1602, Theses XL to XLIII, KGW *4*, pp. 23–4). The Earth's soul is discussed again, at considerable length, in the final chapter of *Harmonices Mundi* Book IV, which is concerned with 'sublunary nature and the soul's inferior faculties, particularly those on which Astrology depends' (HM IV,

Ch. VII, KGW *6*, p. 264). In this final chapter, Kepler gives *Timaeus* as the authority for the existence of a soul of the World (KGW *6*, p. 265, ll. 4–7), but points out that his own theory of the soul of the Earth takes account of the different types of soul described by Aristotle, and ascribes to the Earth no more than the humblest kind of soul, one capable only of recognising geometrical stimuli and responding to them. The Earth's soul is thus postulated to account for a particular group of related phenómena which cannot be accounted for in any other way.

Kepler's geometrical explanation of Aspects is contained in *Harmonices Mundi* Book IV Ch. V. This chapter is set out in mathematical form, as a series of definitions, axioms and propositions, with a small amount of linking text, very much in the manner of *Harmonices Mundi* Books I and II. After defining what is meant by an astrological configuration, and what is meant by describing it as powerful (*efficax*), Kepler proceeds to state two axioms, upon which, he tells us, the whole discussion will depend. They are:

> 'Axiom I
> The arc of the Zodiac cut off by the side of a convex or star polygon which forms congruences and is knowable measures the angle of a powerful Configuration.
> Axiom II
> The angle of a convex or star polygon which forms congruences and is knowable is the measure of the angle of a powerful Configuration.' (KGW *6* p, 241)

These two axioms resemble the axioms which related knowable polygons and Consonances in Book III, Chapter I, the main difference being that the new axioms give not one but two forms of relationship between the physically-realised harmonies and their geometrical prototypes, through the arc cut off by the side and through the angle between two neighbouring sides.

The paragraphs which immediately follow the statement of the axioms, before the first proposition, are concerned with expanding the meaning of the axioms. In particular, Kepler points out that the two axioms in fact lead to the same set of configurations. This is most easily illustrated by bringing together all the diagrams of Aspects which occur later in the chapter, from which we can see that the angle subtended by the side of a congruence-forming and knowable polygon at the centre of its circumcircle is always equal to the angle of another congruence-forming and knowable polygon (for

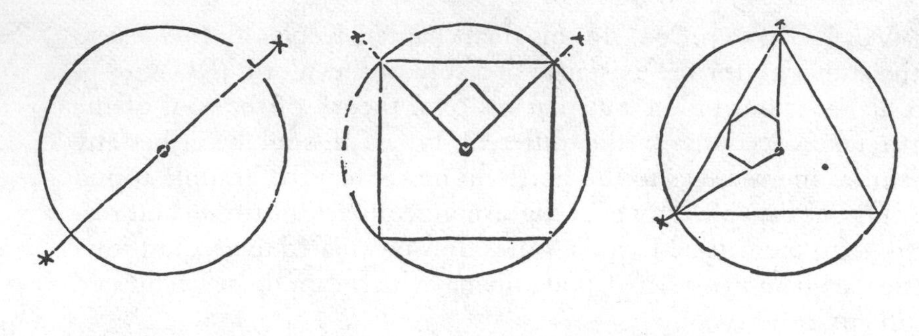

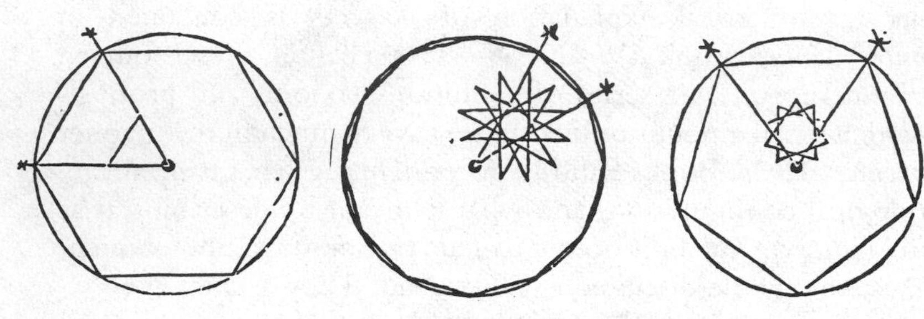

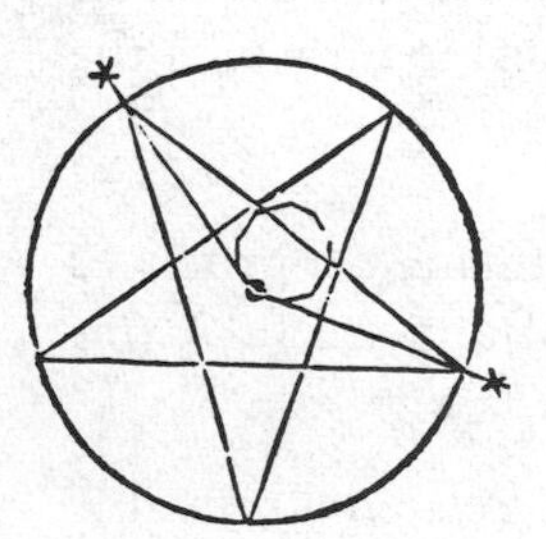

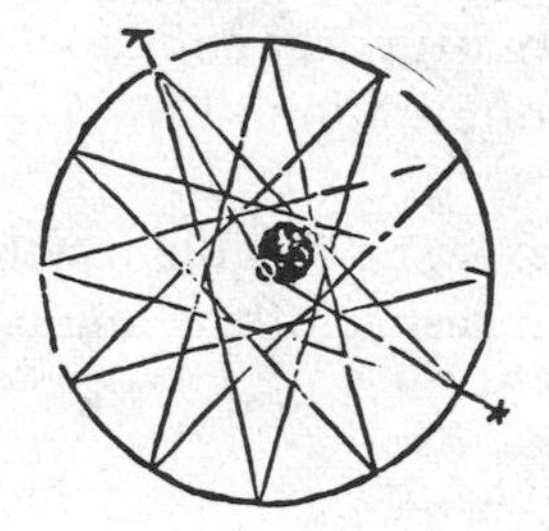

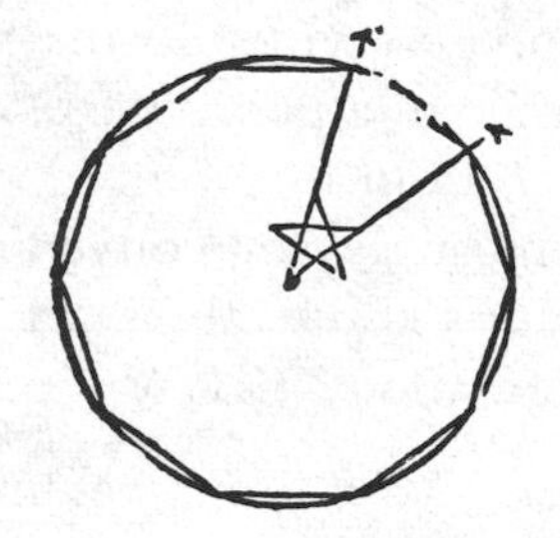

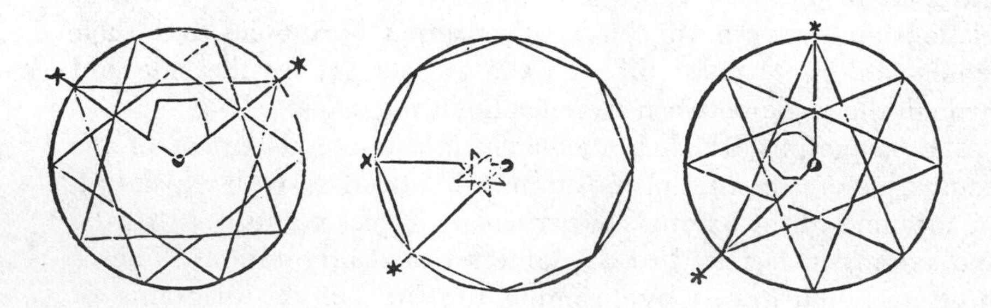

Figure 5.8 Aspects and polygons (extracted from HM IV, Ch. V)

the square it is in fact an angle of the same polygon) (see figure 5.8). Since both axioms give the same set of Aspects it would appear that Kepler has committed the mathematical solecism of employing two axioms where one would apparently have sufficed. However, the use of both axioms has the advantage of allowing him to relate each Aspect to two polygons, the 'central' one and the 'circumferential' one. A decision between the two might have seemed unnecessarily arbitrary at this stage of the proceedings for, as we shall see, Kepler does not seem to regard his axioms as being, by definition, true, but rather as reasonable assumptions that might be susceptible of proof.

The series of propositions which immediately follows the axioms is concerned with establishing the relative importance of the 'central' and 'circumferential' polygons for their corresponding Aspect, and deciding which properties of the polygons should be considered as determining the properties of the Aspects. The first proposition looks back to the Consonances: 'Aspects are more closely connected with the circle and its arcs than the consonances are' (KGW *6*, p. 242, l. 28). This is proved simply by considering the relationship of each to its associated circle:

> 'the consonances do not depend immediately on the circle and its arcs on account of their being circular, but on account of the length of the parts, that is, their proportion one to another, which would be the same if the circle were straightened out into a line. Whereas the Aspects, by definition I, are angles, which the circle measures with its arcs, and in no other way except by remaining what I have called it, that is, by continuing to have a circular shape and to remain complete' (KGW *6*, p. 242, l. 35 ff).

Kepler adds that the Consonances do not always involve the whole circle, but sometimes only ratios of parts of it, whereas Aspects do always concern the whole circle. This mixture of physical and mathematical reasoning is characteristic of most of the propositions, though the relative importance of the two components varies from proposition to proposition. For example, Proposition IV that 'Congruence of figures is more influential than Knowability in making a configuration powerful' (KGW *6*, p. 245) is clearly mainly a matter of physics. The following proposition, that 'Congruence is a property of the Circumferential rather than the Central figure' is entirely mathematical. By way of proof, Kepler asserts, quite reasonably, that the capacity to form congruences is a property of the figure as a whole and the circumferential polygon is employed as a whole (i.e.

the circle goes through all its vertices) whereas the central polygon has only one of its angles at the centre (KGW *6*, p. 245, ll. 35–8). This straightforward mathematical insight provides the justification for using the hierarchy of polygons defined by considering congruences as a means of determining the relative status of Aspects – once Kepler has shown, in Proposition VI, that Aspects are to be seen as determined more by their circumferential than their central polygons, and, in Proposition VII, that for the circumferential polygon congruence-forming is more important than knowability of the side (and vice versa for the central polygon) (KGW *6*, pp. 246–50).

Proposition VIII is a partial converse of Axiom I, namely that the arc of a circle cut off by a figure which does not form congruences does not correspond to an Aspect. It is by no means clear why Kepler did not make this proposition an axiom, unless, perhaps, he thought it too important to be asserted without proof: his comment on it is 'Behold the cause why, although the knowable figures are infinite in number, though of various rank, yet the Aspects are few' (KGW *6*, p. 250, ll. 17–18). However, it is of the nature of axioms to have just this importance, as Kepler had acknowledged after stating his two axioms earlier in this chapter. Moreover, though the two axioms just quoted are called axioms, the series of propositions which follows them does seem to have been at least partly designed to justify accepting them as true, and Kepler's comment on his third axiom (which appears immediately after Proposition VIII and states that the arcs of a circle whose corresponding polygons are of higher rank in congruence and knowability will give more powerful configurations) begins 'If the first two axioms are acceptable (*consentanea*) . . .' (KGW *6*, p. 250, l. 23). It therefore appears that the word 'axiom' does not mean quite the same to Kepler as it does to a modern mathematician, and seems to have done to Archimedes. Kepler's use of the word in this chapter is rather closer to Copernicus' usage in the *Commentariolus*, where 'axiom' apparently means something we shall use as if it were true but would like to prove one day if we can (see Swerdlow 1973). Like Kepler, Copernicus also shows no anxiety to reduce his axioms to the minimum number or to ensure their mutual independence.

Having stated his third, and final, axiom, Kepler proceeds to list the twelve Aspects whose existence can be deduced from the axioms, referring to the results of Books I and II for proof that the polygons involved are congruence-forming and knowable (Prop. IX, KGW *6*, pp. 250–1). The next group of propositions, numbers X to XV, is concerned with establishing the relative degrees of power of the Aspects, on the basis of the nobleness of the corresponding polygons established in Books I and II (i.e. using the criteria set out in Axiom III). Kepler works from the strongest Aspects, opposition and conjunction, which correspond to the diameter of the circle (Prop. X, KGW *6*, p. 251; see figure 5.8a), to the weakest configurations, 'configurations which hesitate between power and powerlessness, namely the 24° arc from the pentekaedecagon and the 18° arc from the icosigon' (Prop. XV, KGW *6*, p. 256). The lowest grade of configurations which are definitely accepted as Aspects is that of the decile, tridecile, octile and trioctile Aspects (Prop. XIV, KGW *6*, p. 254; see figures 5.8 d and e). This enumeration of the effective configurations, in order of decreasing power, brings Kepler's chapter to an end.

His following chapter, Chapter VI, returns to the comparison of astrological and musical harmonies, with a discussion of the fact that there is not the same number of Aspects as of Consonances. As we have seen, Kepler had already discussed the physical distinctions to be made between the two types of harmony (in Chapter IV). The new discussion is concerned with differences in the mathematical formulation, which were dictated by the physical differences, and have led to Aspects which do not correspond exactly with the Consonances described in Book III. Chapter VI thus provides philosophical justification for the mathematical reasoning of Chapter V. Kepler no doubt thought he had better show first that his mathematical method would give the desired result and leave philosophical arguments until later.

The final chapter of *Harmonices Mundi* Book IV, Chapter VII, is, as we have already mentioned, concerned with the soul of the Earth. In neither of these last two chapters is there any discussion of how Kepler's Aspects are to be applied in the practical business of astrology. Indeed, Kepler begins Chapter VI by ruling out such practical considerations: 'On what occasions powerful Configurations are detected, and the increase in their Number [i.e. beyond the conventionally accepted number], this is not the place to recall, for such matters belong to Astrology; and I dealt with them twelve years ago in my book on the new star and the Fiery Trigon [i.e. *De Stella Nova*, Prague, 1606], in Chapters VIII, IX and X . . .' (KGW *6*, p. 257, ll. 4–7). Thus, although the main title page of *Harmonices Mundi Libri V* describes the fourth book as 'metaphysical, psychological and astrological', and the title page of Book IV reads 'On the harmonic configurations of stellar rays at the Earth, and their

144 DE CONFIGURATIONIBUS

CAP. V.

Si duo prima axiomata sunt consentanea vero, erit & hoc: quia propter quod unumquodque est tale; illo intenso, istud etiam magis erit tale. Sic autem intellige; quod in figura circumferentiali prior sit comparatio graduum Congruentiæ, in centrali prior graduum Scibilitatis, denique potiores partes circumferentialis figuræ.

Propositio IX.

Configurationes efficaces sunt, quæ intercipiunt Arcus circuli Zodiaci istos:

Gr. 180. Oppositio ☍, ex Diametro circuli: ut in Fig. I.
Gr. 90. Quadratus □, ex Tetragono: ut in figura II.
Gr. 120. Trinus △. & 60. Sextilis ⚹, ex Trigono & Hexagono, ut in figura III. IV.
Gr. 45. Octilis vel Sequadri, & 135. Trioctilis vel Sesquadri ✱ ex Octogono & Stella ejus: ut in fig. V. VI.
Gr. 30. Semisexti ✳, & 150 Quinquuncis, Ex Dodecagono & Stella ejus: ut in fig. VII. VIII.
Gr. 72. Quintilis ±, & 108. Tridecilis seu Sesquintilis : ex Pentagono & Stella Decagonica: ut in fig. IX. X.
Gr. 144. Biquintilis ✡, & 36. Semiquintilis seu Decilis: ex Stella Pentagonica & Decagono, ut in figura XI. XII.

Quòd hæ figuræ sint Scibiles & demonstrabiles, ostensum est Libro I. quòd & Congruæ, libro II. Quòd verò configurationes expressorum à talibus arcuum sint efficaces, id habent axiomata I. II. præmissa.

Propositio X.

Efficacitatis Aspectuum gradus primus & fortissimus, est Conjunctionis ☌. & Oppositionis ☍.

Characteres usitati.

☍

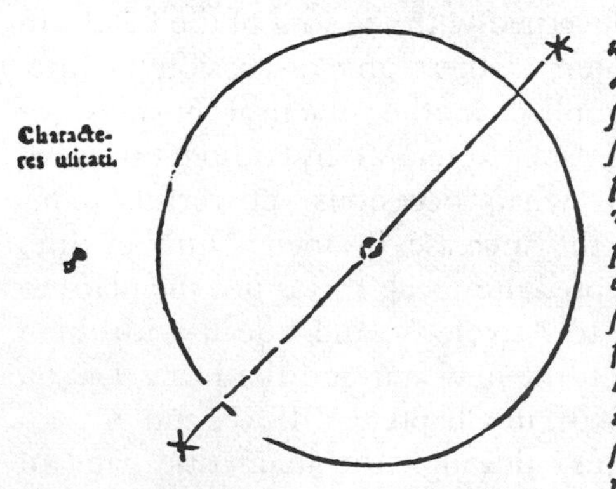

Nam in Conjunctione congruunt radij duo in eandem lineam, & ab eadem plaga descendunt; in Oppositione ☍ à plagis quidem diversis descendentes, nihilò tamen minus fiunt partes unius continuæ lineæ. Hæc verò perfectissima est Congruentia & principium quoddam omnis congruentiæ. Sic cùm conjunctionem repræsentet punctum signatum in circumferentiâ circuli; oppositionem verò, Diameter; hæc certè sunt principia, illa & mensura omnis in hoc genere scientiæ; cùm omnis in circulo lineæ rectæ scientia contineatur determinatione demonstrativâ per Diametri vel longitudinem vel potentiam: ut libro primo

Figure 5.8a Opposition (HM IV, Ch. V)

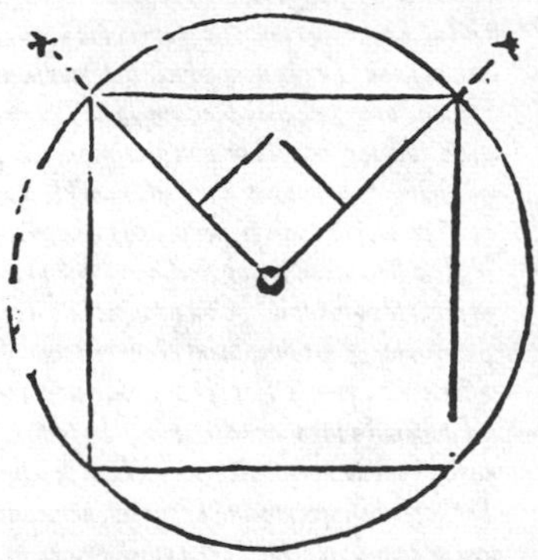

HARMONICIS LIB. IV. 145

CAP. V.

primo patuit. Ergò per Axioma III. principium etiam Efficacitatis in his est Aspectibus.

Propositio XI.

SEcundus in Aspectuum Efficacitate gradus est Quadrati □.

In Quadrato enim concurrunt praerogativae multae, quarum prima, quod similis est centralis figura, circumferentiali: quare quoscunque illa gradus obtinet in Congruentia & Scibilitate, ij quodammodo duplicati intelliguntur, respectu caeterorum Aspectuum. Sicut enim Quadratus primus post Oppositum ab exilitate lineae explicatur in aliquam latitudinem seu amplitudinem superficialem areae Tetragonicae: sic caeteri Aspectus ab identitate figurarum Aspectus Quadrati, discedunt in aliquam figurarum alteritatem. Cùm igitur aliàs in physicis unita virtus sit fortior, erit etiam in hac ideali & objectivâ impressione, major gradus fortitudinis, ubi figurae locis distinctae, altera sc. centralis, altera circumferentialis, specie eadem fuerint. □

Deinde quantum ad Congruentiam, illa in Tetragono perfectissima est & omnivaria, nam secum ipsa congruit haec figura in solido ad cubum formandum, qui mensura est omnis soliditatis, & congruit simplicissimè, ternis tantum angulis ascitis: congruit & in plano secum ipsa, quaternis angulis: congruit rursum in solido cum Trigono, Pentagono, Hexagono, Octogono, Decagono variè, ad formandas figuras solidas, congruit cum ijsdem omnibus, insuperq; & cum Dodecagono & Icosigono quadantenus, ad planitiem sternendam: qua in proprietate illa à nulli aliâ superatur.

Tertiò area Tetragoni est effabilis, quod principium est singularis alicujus & eximiae Congruentiae in plano: ut certus arearum hujus figurae numerus absumat certum quadratorum diametri numerum, & sic figurae non tantùm ipsae inter se angulis & lateribus congruant, sed quodammodò, certis sc. suis lineis, etiam cum quadrati diametri lateribus. In hac proprietate Quadratus aspectus solum Semisextum habet ex parte socium. Vide lib. II.

Quartò, nec ignobilis est gradus scientiae lateris, quod est effabile potentià: quo gradu praecellit caeteris figuris omnibus, excepto Sexangulo: neque tamen illi propterea loco cedit; cùm Scibilitas non sit comparanda Congruentiae, ut explicatum est suprà; & verò valeat accumulatio praerogativarum, ad augendam Efficaciam, per Axioma III. hujus.

Propositio XII.

TErtius Efficacitatis Gradus est Trini △, Sextilis ✶. & Semisexti ⚹.

Quòd Trinum, Sextilem, & Semisextum, in eodem gradu colloco; facis

T

Figure 5.8b Quadrature (HM IV, Ch. V)

146 DE CONFIGURATIONIBUS

CAP. V. ſacit proprietatum non identitas, ſed æquipollentia. Primùm eorum figuræ principales, in congruentia plana tradunt mutuas operas: coëunt n. & inter ſe variè & cum alijs, vt quadrato. Præcellunt quidem hic Trigonus & Hexagonus, quia etiam ſecum ipſæ ſingulæ ſpecies congruunt; præcellit Trigono Hexagonus, quia perfectiſſimam obtinet in plano congruentiam, ſolis ſc. ternis angulis: præcellunt ambo Dodecagono, quia etiam in ſolido congruunt illi cum figuris alijs, quod nō poteſt Dodecagon⁹ At viciſſim præcellit reliquis Dodecagon⁹ effabilitate area, cùm illorū areæ ſint mediales & ſic ignobiliores: quæ arearū differentia, vt jam modo dictum, redundat in congruentiæ perfectionem.

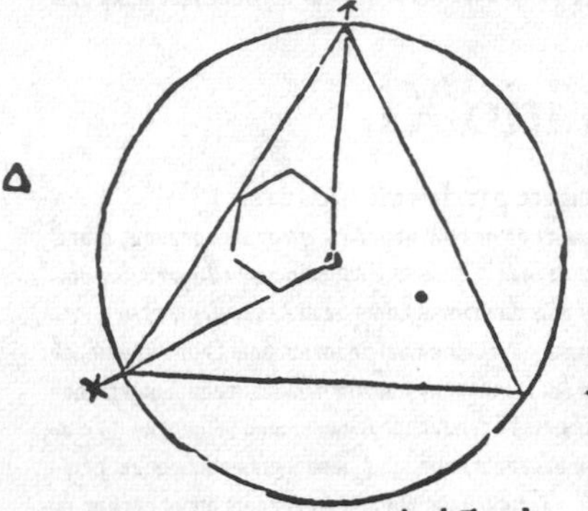

Sic etiam Trigonus præcellit rurſum Hexagono, eo quòd ſecum ipſa Trigonica ſpecies in ſolido congruit variè, gignitq; tria corpora regularia; Hexagon⁹ tantùm cum figuris alijs congruit. Ita penſatis inter ſe diverſarū proprietatū ponderib⁹, Congruentia, quæ primum & præcipuum elementum eſt efficaciæ, penes hos tres propemodum ad æquilibrium perducitur. In ſcibilitate primas tenet Sexangulum, cujus lat⁹ effabile; ſecundas Trigon⁹, occupat enim eundem cum Tetragono gradum, habens latus effabile potentiâ, viliori tamen proportione: ultimus hic eſt Dodecagonus, habens latus ineffabile. Verùm ſcibilitas nec præcipuum eſt ad Efficaciam argumentum, nec in figura præcipuâ, hoc eſt circumferentiali, conſideratur: ſed tantum in centrali min⁹ præcipuâ. Quæ ſi quid poteſt, Trinum paulò reddit efficaciorem Sextili, quia Trinum format angul⁹ Hexagoni in centro; paulò min⁹ utriſq; efficacem Semiſextū, quem metitur angul⁹ Stellæ Dodecagonicæ in centro. Eſt tamē cæteris ſequentib⁹ nobilior ſcientia Semiſexti, quia lat⁹ centralis figuræ, in Ineffabilib⁹ nobiliſſimæ eſt ſpeciei, ſc. Binomiū, & in earū ſubdiviſione duplici, ſemp priores tēet, adeò vt cum ſociâ ſuâ, latere circumferentialis figuræ, rectangulū effabile formet, quod eſt nota perfectionis penè abſolutæ: adeòq; etiam cum Trigono & Hexagono hanc figuram ſcibilitate facit contendere, propter hanc penſationem ineffabilitatis ſuæ, ponderoſam admodum.

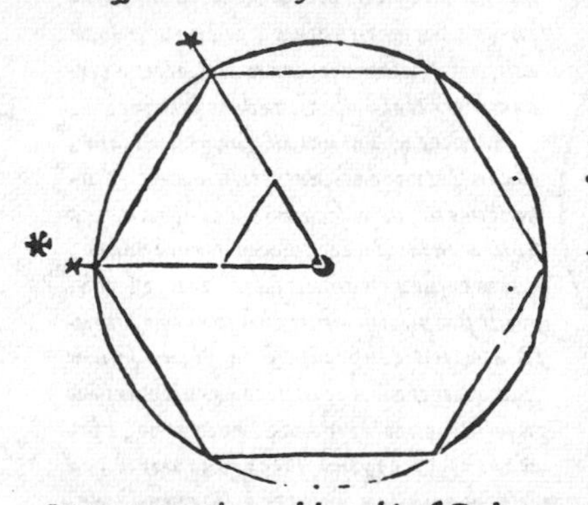

Propo-

Figure 5.8c Trine, sextile, semisextile (HM IV, Ch. V)

148 DE CONFIGURATIONIBUS

CAP. V.

ostendimus pr. IV. majus igitur est, & plus ad efficacitatem potest, creare figuram solidam (quæ est veluti idea quædam mathematica efficacitatis physicæ) quàm latus habere perfectiori gradu scibile. Latus quidem Dodecagoni hoc pacto confert aspectum suum in eandem classem cum subtensis decimæ parti circuli, tribusque decimis: quia contendunt inter se præstantiâ scibilitatis. Nam sicut sociantur inter se duæ illæ subtensæ, fitque minor majoris pars, in proportione divinâ sectionis secundùm extremã & medium: sic etiam latus Dodecagoni & latus ejus stellæ sociantur, & hoc etiam respectu sectionis & compositionis alicujus, non tamen proportionalis. Et hæc quidem biga posterior cadit in primam speciem Ineffabilium, quæ complectitur Binomines & Apotomas; at vicissim illa prior biga acquirit novam proprietatem sectionis secundùm extrema & medium: ut videre est lib. I. Quare non tantùm pensantur hi gradus, sed etiam præcellit nonnihil Decanguli latus. Rectè igitur fecisse, quod Aspectum Quincuncem seu Gr. 150. cum Quintili Gr. 72. & Biquintili 144. eodem gradu locavi, primâ tamen sede his datâ.

Propositio XIV.

QUintus, ultimus & imbecillissimus aspectuum Gradus est Decilis & Tridecilis, Octilis & Trioctilis.

Quintum locum feci Decili & Tridecili (Mæstlinus Semiquintilem, & Sesquiquintilem appellat) quos in Ephemeridibus hactenus omisi: quibus sociavi Octilem & Trioctilem seu Sequadrum, & Sesquadrum: quos Calendariographi ex mea quidem suggestione & nonnullâ Ptolemæi authoritate, sed nimis cito & & inconsideratè arripuerunt. Probandum igitur est utrumque, primùm imbecilliores esse hos quatuor, Quintili & Biquintili; deinde, Decilem & Tridecilem, fortiores Octili & Trioctili, parùm admodùm.

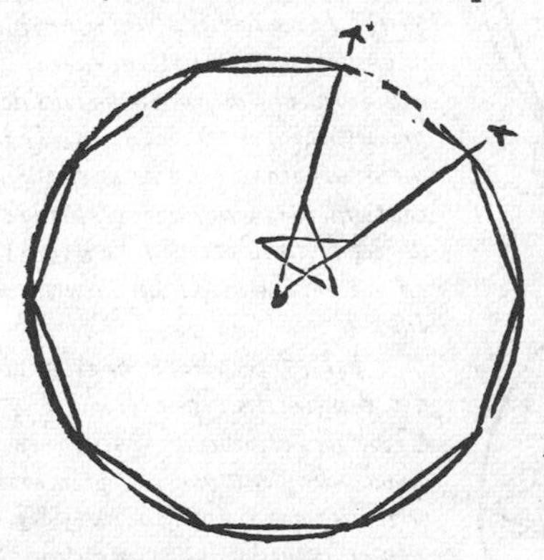

Cùm igitur Propositiones nostræ præcipuum ad Efficaciam momentum collocent in congruentia figuræ præcipuæ, hoc est circumferentialis: manifestum est, Pentagonum & stellam ejus, congruere cum suæ quamque speciei figuris, ad solidum perfectum formandum, ut jam modo dictum; congruere etiam inter se pulchrè ad planum sternendum. Decagonus vicissim & Octogonus cum stellis suis, quæque cum suæ speciei aliis, in solido congruere non possunt. Congruunt quidem, Decagonus & Octogonus, sed cum alijs non omnibus sui generis; stellæ verò inchoant aliquam congruentiam in solido, at non absolvunt: etiam in plano ignobilior est earum congruentia, quia nec mutuas tradunt operas, quævis figura cum suâ stellâ solitariæ, ut Pentagonus cum suâ: sed cum illis suis stellis, & Octogonus cum Tetragono, in societatem veniunt, congruentiæ alie-

Figure 5.8d Decile (HM IV, Ch. V)

tiæ alienæ, illamque, quo minus continuari possit, ipse Decagonus impedit: Stella ejus etiam hiulcam in mediis interceptis spaciis facit congruentiam. Octogonus verò & stella ejus, alternis juvant continuationem congruentiæ, admixtis Quadratis; Congruentia fit diversiformis. Ita penè pares sunt hæ quatuor in congruentiâ planâ; præsertim cùm areas utraque figuræ habeant ineffabiles. At in scibilitate multùm præcellit secta Pentagonica. Primùm si centrales figuras consideremus, quæ sunt hic jam Pentagonus & Stella ejus, illarum quidem latera sub eâdem speciem ineffabilium cadunt, cum lateribus Octogoni & stellæ; existentia Elasson & Meizon: sin circumferentiales, quæ hic sunt latera Decagoni & stellæ ejus: illa non tantum sunt ex specie nobiliori Bimedialium & Apotomarum, cùm Octogonicolæ sint ex quarta specie, quæ est Meizon & Elasson: sed acquirunt etiam omnia latera Pentagonicæ sectæ, nobilissimam proprietatem sectionis secundùm extremam & medium; quæ planè nihil attinet lineas Octogonicas. Quòd si Octogonica secta nonnihil præpollere visa est in Congruentiâ; hic jam vicissim, multò fortiùs deprimitur à Pentagonicâ. Rectè igitur utrasque, ut de præstantia contendentes, in unam classem redegi, præmissâ tamen Pentagonicâ. Consulatur de his identidem liber I.

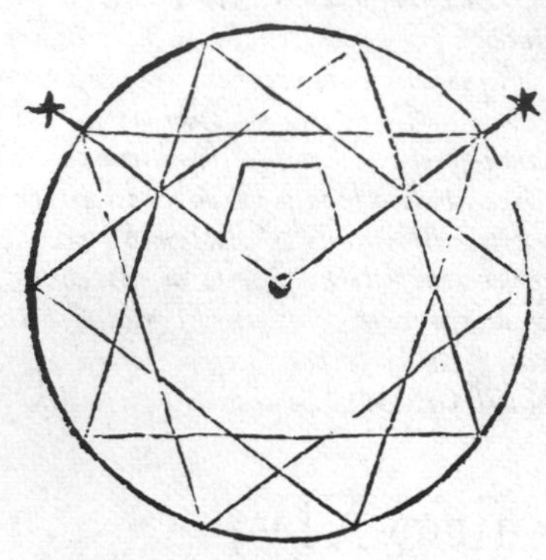

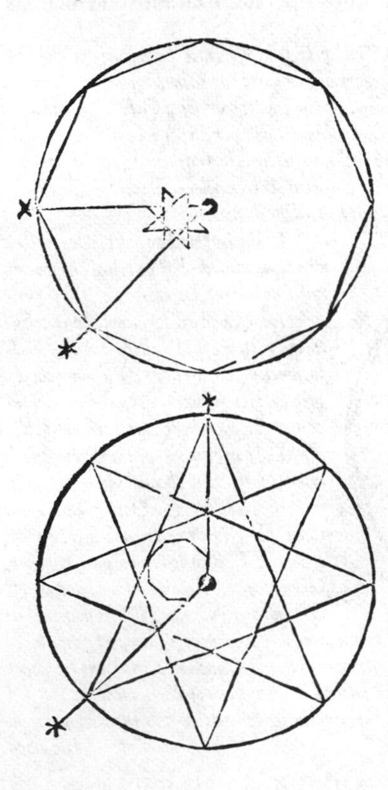

Est & peculiaris prærogativa Biquintilis, præ Tridecili & Trioctili, etiamque Quincunce, quòd stella Pentagonica, primaria scilicet illius figura, aptissimum & Trigonici æmulum habet angulum: quia ut tres anguli Trigoni, sic etiam quinque anguli stellæ Pentagonicæ, junctim utrinque æquantur duobus rectis, ut sic latera angulos formantia, circuli arcus, quæ-

Figure 5.8e Tridecile, octile, trioctile (HM IV, Ch. V)

HARMONICIS LIB. IV. 147

CAP. V.

Propositio XIII.

Quartus in Efficacitate Configurationum Gradus est Quintilis, Biquintilis, & Quincuncis.

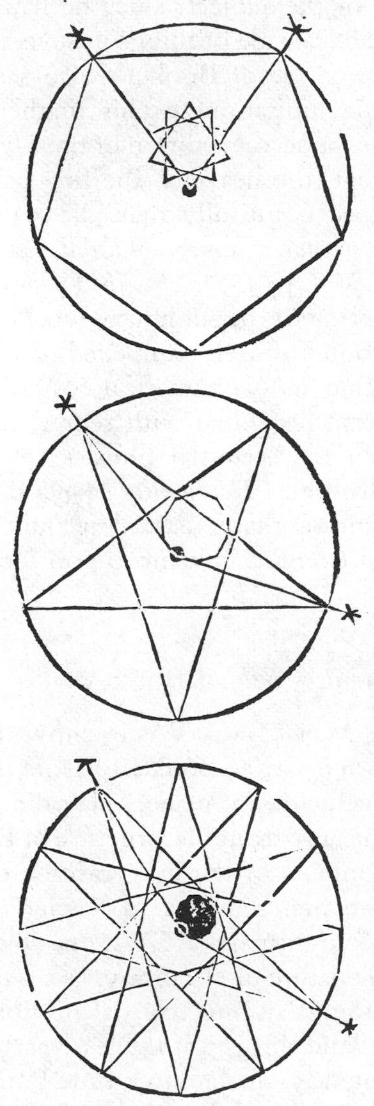

His enim communis est congruentia figurarum primariarum totarum in plano, non tamen singularum specierum secum ipsis; sed primarum duarum inter se mutuò, ultimæ cum alijs sibi cognatis. Præcellunt duo priores aspectus eo, quòd congruunt figuræ, Pentagonus & stella ejus, etiam in solido, faciuntque duas figuras solidas regulares; qua nobilitate penè associant aspectus suos Trino & Quadrato; stella Dodecagonica in solido non congruit. At vicissim præcellit & Dodecagonica, congruentiâ planâ; quam habet continuabilem in infinitum; cùm illæ non longè continuari possint sine mixturâ irregulari. Vide hæc omnia libr. II.

Quod scibilitatem attinet laterum in figuris centralibus; hic etiam medio loco consistunt latera Decagoni & Tridecilis & Dodecagoni, quæ sunt hac in classe centrales, inter latus Trigoni præcedentis, & latera Pentagoni, stellæque Pentagonicæ, centralium figurarum in classe sequenti, Nam libro I. demonstratum est, prius esse in scientia, Decagonicum latus Pentagonico, Tridecile stellari Pentagonico. Itaque & Scibilitas eodem ducit, quo & Congruentia, per pr. VII: quæ hujus potissimùm demonstrationis causâ fuit præmittenda, ne Decilis vel Tridecilis præferrentur Quintili & Biquintili. Si verò quis missâ figurâ centrali, Scibilitatem potiùs in circumferentiali quærere velit, non minus quàm Congruentiam: etsi fatendum est, hoc pacto prælatum iri Decilem Quintili, Tridecilem Biquintili; at meminerit is, præcipuas esse partes Congruentiæ, ut

Figure 5.8f Quintile, biquintile, quincunx (HM IV, Ch. V)

effect on the weather and other natural phenomena', the contents of the book are very little more than a discussion of the theoretical foundations of astrology, rather than what Kepler himself apparently regarded as astrology proper.

Presumably Kepler felt it was not inappropriate for this book to deal only with the foundations of the subject, since he had, as he remarks, dealt with the more practical side of things in several other works, but his restriction of the scope of Book IV also seems to indicate what he took to be the limitations of his mathematical theory. As he said in his notes for the second edition of the *Mysterium Cosmographicum* he had found that 'the heavens, the first of God's works, were laid out much more beautifully than the remaining small and common things' (*Mysterium Cosmographicum*, Frankfurt, 1621, note on the title page, KGW *8*, p. 15, l. 16). In *Timaeus* Plato had given a mathematical description of the sublunary world as well as the celestial, but Kepler had found himself compelled to apply an Aristotelian distinction and confine his mathematical cosmology to the heavens. We may imagine that he did so with regret, since he does not seem to have believed that celestial physics should be distinguished from terrestrial. However, *Harmonices Mundi* Book IV succeeds at least in giving an a priori mathematical account of the means by which celestial bodies exercise influence upon terrestrial ones.

Astronomy: *Harmonices Mundi* Book V

The main concern of *Harmonices Mundi* Book V is to show that the human music described in the latter part of Book III has its celestial counterpart, in which ratios of velocities of planets play the part of musical intervals. This is clear not only from the structure of Book V as a whole but also from the contents of the long cascade of subordinate clauses which make the first sentence of Kepler's introduction to the book a translator's nightmare (KGW *6*, p. 289, ll. 2–19). The sentence ends 'the whole nature of Harmony, as a whole, as explained in Book III, is to be found among celestial motions; not indeed in the way I had thought; and this is not the least part of my joy; but in another very different way, at the same time both most striking and most perfect'. 'The way I had thought' probably refers to the astronomical section of the projected *De Harmonice Mundi* which Kepler had intended to deal with the causes of the periods of the planets (Kepler to Herwart, 14 Dec. 1599, letter 148, KGW *14*, p.

100). Ironically enough, though Kepler abandoned this project, it is for its 'explanation' of the periods of the planets that *Harmonices Mundi* Book V is now best known. For Kepler's third law is, in Kepler's own terms, to be seen as an explanation of the periods, since it establishes their dependence on the mean radii of the planetary orbs, and to Kepler, as we have seen, the dimensions of the orbs were a more fundamental characteristic than the periods (see Chapter III above).

Kepler did not discover his third law until shortly before he finished writing Book V, while the first books of the work were already being printed (see Haase, 1971, and Gingerich, 1975). In *Harmonices Mundi* Book V, Chapter III, Kepler gives his own account of the slightly confused final stages of his discovery of the third law:

> '. . . I thought of it on 8 March of this year 1618, but was unsuccessful in my attempt to verify it numerically, and rejected it as untrue; then on 15 May it came back to me and with a new assault drove the obscurity out of my mind; seventeen years' work with Tycho's observations and my present meditations on this subject were in such good agreement that at first I thought I was dreaming and was assuming the truth of what I was trying to find. But it is most certain and most exact that the proportion between the periods of any two planets is precisely three halves the proportion of the mean distances, that is of their orbs' (KGW *6*, p. 302, ll. 14–23).

Kepler then goes on to explain more precisely which dimension of the orb is to be taken as the 'mean distance'.

The writing of *Harmonices Mundi* Book V was finished twelve days after this discovery, on 27 May 1618, and the modifications to Book V occasioned by the discovery were eventually made on 19 February 1619 (HM V, Ch. X, KGW *6*, p. 368). The new law does not play a large part in the astronomy of Book V, and there is, in fact, no reason why it should. It clearly cannot affect the theory that the dimensions of the planetary orbs are determined by the Platonic solids: because this theory is concerned with the inner and outer surfaces of the orbs and not with their mean radii as the third law is. Moreover, the third law is equally irrelevant to the main astronomical concern of *Harmonices Mundi* Book V, namely the musical ratios to be found among the velocities of the planets at extreme points of their orbits, since the law is concerned only with the periods of the planets, that is, with their mean motions, not their motion at particular points. However, the third law plays an important part in the cosmological work of Book V, the deduction of the dimensions of the planetary spheres from the 'harmonic' archetype described in the penultimate chapter of the book (HM V, Ch. IX, see below).

The third law is not given a prominent position in *Harmonices Mundi* Book V, but it is not submerged in calculations as the first two laws were submerged in the *Astronomia Nova*. Caspar has generously included the law in his *Personenregister*, under the name '*Keplers drittes Planetengesetz*', but it is, really, where one would expect to find it: in the chapter which summarises the astronomical results which will be required in the later part of the work. This astronomical summary forms the third chapter of Book V, the first chapter having been devoted to an account of the five Platonic solids and other geometrical matters (including a brief description of one of Kepler's star polyhedra). The second chapter of Book V discussed the connection of harmonic proportions with 'the five regular figures' i.e. the Platonic solids (Kepler knows his star polyhedra should 'by rights be accepted as regular', but continues to use the word 'regular' as if it singled out the Platonic solids – see Field 1979a, Part I). This numerological chapter may at first appear to be a more sophisticated version of the numerological chapter of the *Mysterium Cosmographicum* (Chapter X). However, Kepler's notes on that chapter in the second edition refer the reader to *Harmonices Mundi* Books I and II for geometrical alternatives to the numerology and not to this numerological chapter in Book V. The reason for this seems to be that in the *Mysterium Cosmographicum* Kepler was trying to use the properties of the Platonic solids as an explanation for properties of numbers, a role which was, indeed, taken over in the later work by regular polygons (hence the reference to *Harmonices Mundi* Books I and II), whereas in *Harmonices Mundi* Book V Chapter II he is merely showing that some harmonic proportions, whose geometrical origins have already been established, are observed to occur among the geometrical properties of the solids. He is thus concerned with a mere connection not a determining cause.

As we saw in Chapter IV above, the astronomical summary in *Harmonices Mundi* Book V Chapter III includes a qualitative account of the agreement between the dimensions of the planetary orbs calculated by considering the circumspheres and inspheres of the Platonic solids and the dimensions found by calculation from Tycho's observations, after which Kepler notes that the agreement is not exact and concludes

> 'From this it appears that the exact proportion of the distances of the planets from the Sun was not taken from the five regular solid figures alone; for the Creator does not depart from his Archetype, the Creator being the true source of Geometry and, as Plato wrote, always engaged in the practice of Geometry' (HM V, Ch. III, KGW *6*, p. 299, ll. 29–32).

Though Kepler does not quote any numerical values for the dimensions of the orbs, it is certain that he must have calculated values like those shown in table 4.4 (apart from the column of percentage errors). Now, the absolute errors in that table do not show a marked improvement over the errors in the corresponding figures as calculated on the basis of the much less accurate observational data available to Kepler at the time he wrote the *Mysterium Cosmographicum* (see table 4.5). Kepler is justly famous for having made crucial use of an estimate of observational error to reject a theory, but the present case is a rather more complicated one. It would seem that he had no practicable way of finding out how Tycho's observational errors might affect the 'observed' radii of the planetary spheres. One suspects he made a crude comparison of the absolute disagreement between theoretical and 'observed' values and decided that the agreement should have been improved more than it seemed to have been. The very stringent standard implied by the words 'for the Creator does not depart from his Archetype' may perhaps have been applied later, when Kepler had found a 'harmonic' explanation of the proportions of the orbits which fitted closely with what was observed. The later theory created its own standard of exactness, as well as explaining why the earlier one had failed to meet it. Kepler's belief that such a standard could be met must also have been bolstered by his discovery of the third law, whose agreement with observation was very close indeed (see Gingerich, 1975).

Having pointed out that the archetype proposed in the *Mysterium Cosmographicum* is inadequate because it does not give an exact account of the 'observed' internal and external radii of the orbs, Kepler immediately begins to discuss the thickness of the orbs. As we have seen, the theory in the *Mysterium Cosmographicum* made no attempt to explain the different thicknesses of the various planetary orbs. Considering the thickness of the orb, that is, the eccentricity of the orbit of the planet, naturally leads one to consider the variable speed at which the planet moves round its orbit. This progression must have seemed even more natural to Kepler, who had spent many months on the calculations which eventually related the orbital velocity to the length of the radius vector by means of the Area Law, his first success in his 'battle with Mars'. It thus seems possible that the idea of the musical archetype described in *Harmonices Mundi* Book V may be the product of a line of thought such as that pursued in Book V, Chapter III, where Kepler starts by noting a philosophical inadequacy in the earlier archetype, namely its failure to deal with the thickness of the orbs, and goes on to examine the changes in speed of the planet and thus its extreme speeds, at perihelion and aphelion (HM V, Ch. III, p. 299, l. 32 to p. 300, l. 30). In any case, Kepler merely passes from his note of the inadequacy of the earlier archetype to a discussion of planetary velocities (taking in the third law on the way) and then, in the next chapter, goes on to search for musical ratios among various possible measures of the orbital velocities. There is no description of the discovery of the musical archetype to serve as pendant to the description of the discovery of the polyhedral one in the preface to the *Mysterium Cosmographicum*.

As we have seen, in *Harmonices Mundi* Book IV Kepler was dealing with harmonious relationships among the apparent positions of the planets, as seen from the Earth. He makes it clear from the very beginning of Book V Chapter IV that his concern is not with apparent properties but with absolute ones, that is, properties measured with respect to the still centre of the Universe, the Sun (or, in the Tychonic system, the centre at least of planetary orbits):

> 'So, having disposed of the fantasy of retrogradations and stations, and having got to the heart of the matter, the individual motions of the planets in their true eccentric orbits (*in suis genuinis Orbitis eccentricis*), there now remain the following properties that distinguish the planets one from another: 1. Distances from the Sun, 2. Periodic times, 3. Daily orbital arc on the eccentric [i.e. as an angle], 4. Length of daily path along these arcs, 5. Angles at the Sun or the apparent arc traversed in a day as seen from the Sun' (KGW *6*, p. 306, ll. 5–10).[21]

The wording of the fourth item is none too clear – and Caspar's translation of it appears to be meaningless (Kepler trans Caspar, 1973, p. 295). However, Kepler's later discussion indicates that the third item refers to arcs of the orbit seen as angular arcs traversed in the planetary sphere, and lends support to the above translation of the fourth item. Kepler says

'. . . indeed, arcs of two eccentrics, equal or denoted by the same number, nevertheless have different true lengths in relation to the complete eccentrics. For example, one degree in the sphere of Saturn is about twice as large as one degree in the sphere of Jupiter. And again, the daily arcs on the eccentric expressed in astronomical numbers [i.e. as angles] do not show the proportion between the true distances the globes traverse through the aether, because single units denote a larger portion of distance in the wider circle of a higher planet, and a smaller one in the narrower circle of a lower one' (KGW *6*, p. 306, l. 32 to p. 307, l. 3).

This explanation seems to indicate that items three and four in Kepler's list are in fact concerned with true motions, that is, motions referred to the Sun, and their use of the terrestrial day as a unit of time is merely conventional (and convenient). Apart from the period, all the items on this list change in magnitude as the planet moves round its orbit. Kepler decides to use the most nearly static values, that is values at the points of the orbits where they change most slowly, namely perihelion and aphelion (KGW *6*, p. 306, ll. 10–25).

He considers the periods first, presumably because they are the simplest to deal with, and gives a table to show that they do not present musical proportions (KGW *6*, p. 307). The distances from the Sun are considered next, and there is another table, showing the aphelion and perihelion distances of each planet, together with the ratio between these distances for each individual planet and the ratio between values of different types for neighbouring pairs of planets, for example, the ratio of the aphelion distance of Jupiter to the perihelion distance of Mars (KGW *6*, p. 309). The only individual planets to yield ratios close to musical ones are Mars and Mercury, but some of the pairs of neighbouring planets also give such ratios. Kepler makes it clear that he is looking for more convincing results than this (KGW *6*, p. 310, l. 1 ff). He turns next to the daily motions, again considering aphelion and perihelion values, in degrees and minutes of arc as seen from the Sun, and in terms of the actual distance traversed, the unit of distance being chosen so that the mean distance of the Earth from the Sun is 1000 (KGW *6*, p. 311). He deals with the linear distances quite briskly, correctly pointing out that for an individual planet the ratio will be the inverse of that obtained from the distances from the Sun in the last table, and adding that it turns out that we do not find musical ratios from the pairs of neighbouring planets either (KGW *6*, p. 311, ll. 16–22).

The motions in arc are compared in the next table, ratios again being calculated as they were for the distances, so that the arcs for each planet are compared with one another and each arc is also compared with the arc of the opposite kind in neighbouring pairs of planets (KGW *6*, p. 312). This table is shown in figure 5.9. The ratios it derives are crucial to the remainder of *Harmonices Mundi* Book V.

The fourth column of the table gives the 'observed' daily motion, as it would be seen from the Sun, in minutes and seconds of arc. The sixth gives the musical ratios which fit most closely the ratios of the extreme speeds of the planets. The fifth column shows Kepler's method of indicating how closely the musical ratios fit: it contains daily arcs which would give this musical ratio exactly. The suggested modifications of the daily arcs in order to obtain perfect musical ratios are all quite small. If the modified values are taken to be correct, then the 'observed' values would have to be assumed to be in error by 1% for the Earth and Venus, by 2% for Saturn and Jupiter, and by 3% for Mars and Mercury. A little later in this chapter, Kepler points out that one can also find a musical ratio in the motion of the Moon, whose extreme hourly motions in arc are 26′26″ (at apogee) and 35′12″ (at perigee). If the latter figure is increased by 3″, the ratio between them is exactly 3 : 4, the ratio for a fourth, 'a

Harmoniae binorum		*Apparentes diurni*	diurni. *Prim Sec.*	*Harmoniae singulorum propria* *Prim. Sec.*	
Diver.	*Conv.*	♄ Aphelius	1.46.a.	Inter 1.48	est $\frac{4}{5}$ Tertia major.
a 1	b 1	Perihelius	2.15.b.	& 2.15.	
d 3	c 2				
c 8	d 1	♃ Aphelius	4.30.c.	Inter 4.35.	est $\frac{5}{6}$ Tertia minor.
		Perihelius	5.30.d.	& 5.30.	
f 1	e 5	♂ Aphelius	26.14.e.	Inter 25.21.	est $\frac{2}{3}$ Diapente
e 5	f 2	Perihelius	3.81.f.	& 38.1.	
h 12	g 3	Tel. Aphelius	57.3.g	Inter 57.28.	est $\frac{15}{16}$ Semitonii
g 3	h 5	Perihelius	61.18.h	& 61.18.	
k 5	i 8	♀ Aphelius	94.50.i.	Inter 94.50.	est $\frac{24}{25}$ Diesis
i 1	k 3	Perihelius	97.37.k.	& 98.47.	
m 4	l 5	☿ Aphelius	164.0.l.	Inter 164.0.	est $\frac{5}{12}$ Diapason cum tertia minore
		Perihelius	384.0.m.	& 394.0.	

Figure 5.9 Motions in arc, seen from the Sun (HM V, Ch. IV)

harmony which is not found anywhere else among the apparent motions' (KGW *6*, p. 313, ll. 37–8). The suggested modification is less than 0.2%.

The first two columns of Kepler's table show the musical ratios in the extreme speeds of neighbouring planets (*motus diversi* when the aphelion velocity of the higher planet is compared with the perihelion velocity of the lower one, *motus conversi* when the perihelion velocity of the higher planet is compared with the aphelion velocity of the lower one). There is no attempt to show in the table how accurately these first two columns represent the 'observed' ratios.

Kepler's own analysis of his table takes the form of considering the ratio by which the 'observed' ratio differs from the nearby musical one:

> 'The proportions of the individual apparent motions [i.e. the motions seen from the Sun] are very close to musical ones; so that Saturn and Jupiter make only a little more than major and minor thirds, the excess being 53 to 54 in the former case and 54 to 55 in the latter, or less, that is about half a comma; . . .' (KGW *6*, p. 313, ll. 19–23).

His analysis of the 'diverse' and 'converse' ratios is made in similar terms. This is appropriate enough, since the ratios which express the differences would correspond to a small interval in pitch if we were concerned with audible music. It appears that this analysis of the slight discordances by means of ratios did not suggest to Kepler that he should consider relative errors in other contexts also. However, calculations show that Kepler's correcting ratios correspond to percentage errors of 1% for the Earth, 2% for Saturn and Jupiter, and 3% for Mars, Venus and Mercury. The corrections to 'diverse' and 'converse' ratios are of the same order.

Kepler's observational evidence for these 'musical' commensurabilities is as good as most of the evidence which has sent modern astronomers in search of explanations for the commensurabilities they have found. Unfortunately, all the modern work seems to have been concerned with commensurabilities in mean motion, usually of systems of satellites. Commensurability in mean motion is, of course, equivalent to commensurability in period, but it does not show up strongly in the periods of the planets (where, as we have seen, Kepler failed to find it). The observed commensurabilites have been the subject of various individually convincing but mutually incompatible explanations, like much else in astronomy. The facts and the explanations are summarised by Nieto (1972) in his book on the Titius-Bode Law and by King-Hele (1975) in a paper read to the Royal Astronomical Society in 1971. They are expounded in more detail by Goldreich (1965) and by Ovenden (1975). Ovenden's paper was written specially for the Kepler volume of *Vistas in Astronomy* and though it does not deal with the matter directly, it leaves the impression that the commensurabilities in extreme motions which Kepler describes might perhaps be the product of secular evolution of the system of orbits along the lines indicated by Ovenden: he shows that the evolution is slowest when the mean motions are close to commensurabilites, so a system is most likely to be observed in this state. More recent work, by Message (1982), shows that if a planetary system starts off close to resonance it will stay close to resonance indefinitely. This is as near as one can reasonably expect to come to being given scientific permission to believe in the validity of Kepler's musical commensurabilities.

We have already noted that Kepler has shown very good observational evidence for the existence of 'musical' ratios among the extreme velocities of the planets. He now proceeds to treat these ratios in musical terms, but acknowledges their observational origin by referring back to the actual angular velocities to check that his musical manipulations have not led him to depart from a reasonable standard of agreement with observation. We shall see below that he does indeed maintain such a standard, but it must be borne in mind that the possibility of setting it up at all depends upon the observational evidence shown in figure 5.9.

On the basis of this evidence, Kepler proceeds to construct musical scales which include notes corresponding to all the pitches associated with the planets. In order to do this, he has to set a definite pitch for the slowest motion and then put all the other motions at the appropriate intervals above it, but since he is interested, first, merely in obtaining every note of the scales, he decides to ignore octaves, so that, for example, the pitch of Mercury at aphelion is brought down six octaves to form the fifth note of a scale, the pitch of Mars at perihelion is brought down four octaves to form the fourth note of the same scale and its pitch at aphelion is brought down three octaves to form the seventh note (see figure 5.11 below). Kepler gives a table to show how the notes may all be brought into the compass of a single octave (see figure 5.10). The plus and minus signs at the right of the table indicate whether the 'observed' motion shown in the table reproduced in figure 5.9 is to be obtained by adding to the value in the new table or by substracting from it. However, the modifications are very small, as can be seen from table 5.1.

Table 5.1 *Errors introduced when bringing notes within octave*. (Compare figure 5.10)

		motion decreased by *n* octaves		motion min	sec	corr. motion	'obs' motion	u–v	$\frac{u-v}{u}$ %
		n	2^n			v	u		
☿	Peri.	7	128	3	0	384′	384′	0	0
	Aph.	6	64	2	34	164′16″	164′	−16	−0.2
♀	Peri.	5	32	3	03	97′36″	97′37″	+1	+0.02
	Aph.	5	32	2	58	94′56″	94′50″	−6	−0.1
⊕	Peri.	5	32	1	55	61′20″	61′18″	−2	−0.05
	Aph.	5	32	1	47	57′04″	57′03″	−1	−0.03
♂	Peri.	4	16	2	23	38′08″	38′01″	−7	−0.3
	Aph.	3	8	3	17	26′16″	26′14″	−2	−0.1
♃	Peri.	1	2	2	45	5′30″	5′30″	0	0
	Aph.	1	2	2	15	4′30″	4′30″	0	0
♄	Peri.	0		2	15	2′15″	2′15″	0	0
	Aph.	0		1	46	1′46″	1′46″	0	0

In this table, the first five columns contain the information given in Kepler's table (figure 5.10), the sixth contains the arc which would correspond to the given motion multiplied by the appropriate factor to restore it to its correct 'pitch', the seventh contains the 'observed' motions (from figure 5.9), the eighth shows the angle that must be added to the new motion to obtain the 'observed' one, and the final column shows the percentage change involved. Even for the largest absolute change, 16″ for the perihelion motion of Mercury, the percentage change is only 0.2%.

Kepler next proceeds to construct two scales (shown in figures 5.11 and 5.12). As can be seen from these diagrams, one note is missing in each scale, A in the first and f in the second. Kepler points out that A was also missing from the ratios described in Book III, Chapter II (KGW *6*, p. 319, ll. 4–5) but makes no comment on the missing f except to note that it is missing (KGW *6*, p. 319, l. 22). Presumably the reason why these gaps do not worry Kepler is that he knows that they will be filled in once he considers the complete compass of each planet, which he does in his next chapter. The scale shown in figure 5.11 is

Motus	*Prim.*	*Sec.*
Perihelij ☿ ſeptimùm ſubdupla, ſeu 128 *va*	3.	0.
Aphelij ☿ ſextùm ſubdupla, ſeu 64*ta*	2.	34.-
Perihelij ♀ quintùm ſubdupla, ſeu 32 *da*	3.	3. +
Aphelij Veneris quintùm ſubdupla, ſeu 32 *da*	2	58.-
Perihelij Terræ quintùm ſubdupla, ſeu 32 *da*	1.	55.-
Aphelij Terræ quintùm ſubdupla, ſeu 32 *da*	1.	47.-
Perihelij Martis quartùm ſubdupla, ſeu 16 *da*	2.	23.-
Aphelij Martis tertiùm ſubdupla, ſeu 8 *va*	3.	17.-
Perihelij Jovis ſubdupla	2.	45.
Aphelij Iovis ſubdupla,	2.	15.
Perihelius Saturni	2.	15.
Aphelius Saturni	1.	46.

Figure 5.10 Notes within the octave (HM V, Ch. V, p. 203).

durus and uses the aphelion motion of Saturn as its lowest note, G. It takes in all the notes of the planets except those associated with the perihelion motions of the Earth and Venus. An adjustment of 4″ has to be made to fit in the aphelion motion of Mercury (but this corresponds to a change of less than 2%). Kepler's second scale, shown in figure 5.12, is *mollis* and starts from the perihelion motion of Saturn, as G, taking in all the notes of the planets. The motions of the Earth have to be taken up an octave to fit into this scale and small adjustments have to be made to all but one of the motions, the exception being the perihelion motion of Mercury. The largest absolute change is 6″, in the aphelion motion of

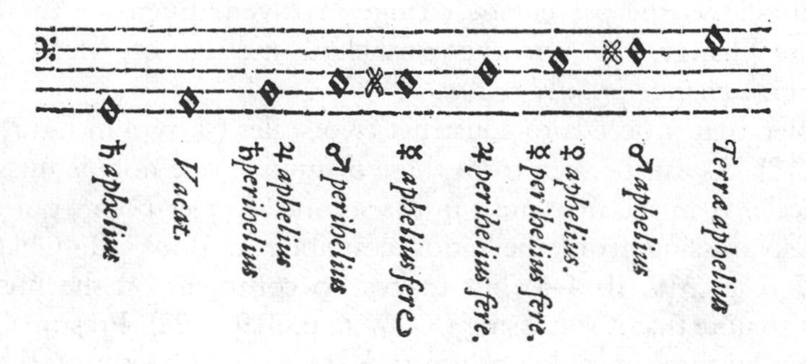

Figure 5.11 Scale in *cantus durus* (HM V, Ch. V, p. 204). The symbol ✻ indicates sharp.

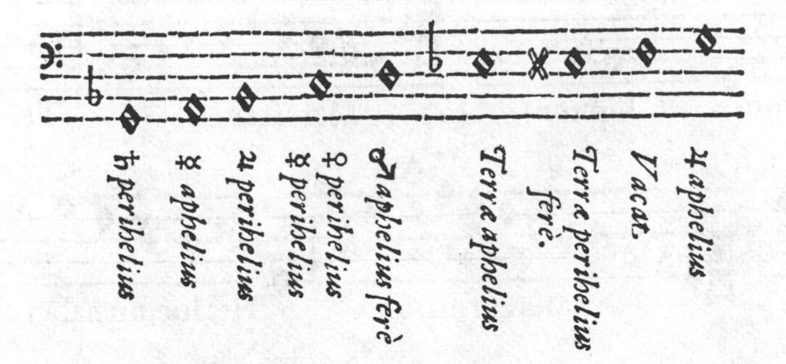

Figure 5.12 Scale in *cantus mollis* (HM V, Ch. V, p. 204). The symbols ✕ and ♭ indicate sharp and flat.

Mars, a change which also gives the largest percentage difference, 3%. All the other changes are less than 2%.

The fact that such scales can be constructed follows, of course, from the commensurabilities Kepler showed in the final table of his previous chapter (figure 5.9). However, it seems decidedly significant that although in that table Kepler showed some adjusted motions which would give exact commensurabilities, he returns to the 'observed' motions for the purposes of constructing his musical scales. He clearly wished to keep as close to the observations as possible.

The following three chapters of *Harmonices Mundi* Book V are concerned with purely musical developments. In Chapter VI Kepler gives the compass of notes corresponding to the range in speed of each planet as it travels round its orbit: Mercury has the largest compass, since it has the largest eccentricity, while Venus is confined to one note and the Earth has a compass of only a semitone (see figure 5.13). In Chapter VII Kepler considers chords produced by all six planets together, and in the very brief Chapter VIII he gives each planet a particular musical voice: 'The properties usually associated with the Bass, mentioned in Book III Chapter XVI, are given to Saturn and Jupiter in the heavens, those of the Tenor we find in Mars, those of the Alto occur in the Earth and Venus, and the Descant's properties belong to Mercury . . .' (KGW *6*, p. 329, ll. 15–19).

Astronomy returns in Chapter IX, which, according to its title, is concerned with 'the origin of the eccentricities of individual planets by harmonies being obtained among their motions' (KGW *6*, p. 330). In fact, after calculating the eccentricities Kepler goes on to calculate the dimensions of the orbs, using his third law. (Perhaps the

Figure 5.13 Compasses of the planets (HM V, Ch. VI, p. 207)

title of Chapter IX was not altered after the discovery of the third law?).

Harmonies are the main subject of Chapter IX, and its first section is an axiom stating that the harmonies must be of as many kinds as possible 'so that this variety may make the world more beautiful' (KGW *6*, p. 331, l. 16). However, the next section is an axiom stating that the regular polyhedra also have a part to play: 'It is proper that the five intervals between the six orbs should to some extent correspond in size with the proportions of the geometrical orbs inscribed in and circumscribed about the five regular solids, and should do so in the order which is natural to the five figures' (HM V, Ch. IX, sect. II, KGW *6*, p. 331, ll. 18–21).

The first series of propositions, which is interspersed with an occasional extra axiom, is concerned with giving a mathematical defence of the association of each of the 'observed' musical ratios with the particular planet or pair of planets for which it has been observed. The procedure is like Kepler's defence of his polyhedral archetype in Chapters III to VIII of the *Mysterium Cosmographicum*, where he proved, for instance, that it was suitable that the cube should be situated between the orbs of Saturn and of Jupiter (see Chapter III above). The justification of the musical ratios for individual planets is always by an appeal to the observed motions: for example, in section XIV, where it is required to show that 'the proportion between the extreme motions of Mars had to be more than a fourth, 3 : 4, and about 18 : 25', the proof depends on the harmonies found between Mars and Jupiter (KGW *6*, p. 336, ll. 26–34). However, the justifications of the musical ratios found between motions of neighbouring pairs of planets generally depend upon the polyhedron associated with their orbs. For example, for the proposition that 'Venus and Mercury had to have a Major Harmony, 1 : 4 a double octave', the proof begins

> 'For just as the Cube is the first of the primary figures, so the Octahedron is the first of the secondary ones, by Chapter I of this book. And just as the Cube, considered geometrically is more outer and the Octahedron more inner, that is, inscribable within it;[22] so, too, in the World Saturn and Jupiter are the first of the superior outer planets, starting from outside; while Mercury and Venus are the first of the inner ones, starting from inside, and between their paths there is the Octahedron, see Chapter III. So Mercury and Venus must have a harmony which is also primary and related to the Octahedron . . .' (KGW *6*, p. 334, ll. 29–35).

This set of axioms and propositions, concerned with the harmonic ratios identified in Book V, Chapter IV, has the heading *Rationes Priores*. The following series, headed *Posteriores Rationes*, is concerned with the scales developed in Chapters V and VI, though the propositions and their proofs do not take a markedly different form from those of the previous series.

The regular polyhedra reappear in their own right in section XLVI, an axiom which explains the words 'to some extent' used in the axiom of section II. This new axiom states that, but for the harmonic causes which have just been described, the geometrical solids would determine the relative sizes of the planetary orbs. The following section, a proposition, describes a modified version of the polyhedral archetype proposed in the *Mysterium Cosmographicum*:

> 'If the inscription of the figures between the planets could be made freely, the angles of the Tetrahedron would touch the perihelion sphere of Jupiter exactly, above, while below, the centres of its faces would touch the aphelion sphere of Mars exactly. The Cube and the Octahedron, with their angles fitted to the perihelion spheres of their planets, would penetrate the orb below them, so that the centres of their faces would lie between its aphelion and perihelion spheres. On the other hand, the Dodecahedron and Icosahedron, fixed to the perihelion spheres of their planets on the outside, would not quite reach the aphelion spheres of the planets inside them with the centres of their faces. Finally, the dodecahedral hedgehog (*echinus dodecaëdricus*),[23] standing with its angles on the perihelion sphere of Mars, would, with the centres

of the parts of its sides contained between two of its points, come close to the aphelion sphere of Venus' (HM V, Ch. IX, sect. XLVII, KGW *6*, p. 354, ll. 19–29).

This is not as neat as the scheme described in the *Mysterium Cosmographicum* but it gives, as we shall see, a rather better fit to the observations. For the present, Kepler is only concerned to establish its mathematical reasonableness, in about a page of geometrical argument that closely resembles the corresponding sections of the *Mysterium Cosmographicum.*

However, whereas in the *Mysterium Cosmographicum* the polyhedral archetype was seen as the only formal cause it is now subject to other factors, as Kepler explains in his next section: 'The inscription of the regular solid figures between the planetary orbs was not entirely free of constraint, for it was hindered a little by the harmonies established among the extreme motions' (sect. XLVIII, KGW *6*, p. 356, ll. 9–11). He proves this proposition by deducing the structure of the Solar system from the harmonies observed among the extreme velocities of the planets, displaying his results in the form of a table at every stage of the calculation, and including intermediate working values in these tables. It is as if he were aware his readers might feel they were watching a conjuring trick.

The first table shows how the eccentricities of the orbits can be deduced from the ratio of the extreme motions (figure 5.14). In the first column of figures Kepler has given numbers whose ratio is

Planetis.	Motuum proportiones	Horum Radices aut prolongatæ aut multiplicium	Ergò ſemidiameter Orbis.	Eccentricitas.	In dimenſione ſemidiametri Orbis 100000
Saturno	64	80			
per XXXVIII.	81	90	85	5	5882
Jovi	6582	81000			
per XXXVIII.	8000	89444	85222	4222	4954
Marti	25	50			
per XI.L	36	60	55	5	9091
Telluri	2916	93531			
per XXVIII.	3125	96825	95178	1647	1730
Veneri	243	9859			
per XXVIII.	250	10000	99295	705	710
Mercurio	5	63250			
per XLIV.	12	98000	80624	17375	21551

Figure 5.14 Ratios of extreme motions of each planet (HM V, Ch. IX, p. 238)

indicated as being the same as the ratio of the extreme motions of the planet concerned. Table 5.2 compares the ratios of the pairs of numbers in Kepler's new table with the ratios of the extreme motions given in Book V, Chapter IV (KGW *6*, p. 312, see figure 5.9). One must presume that it was from these extreme motions that Kepler obtained the ratios shown in figure 5.14, and it is clear from the very small percentage errors that he has done no violence to his observational data in obtaining the numbers in the new table, simple though some of them are. The second column of figures in Kepler's table contains the square roots of the figures in the previous column.

Table 5.2 *Errors in ratios of extreme motions of planets*. (Compare figure 5.14.)

planet	ratio of motions	ap./peri. from ratio rr	ap./peri. 'obs' ro	$\frac{rr-ro}{ro}$ %
♄	$\frac{64}{81}$	.790 123	.785 185	0.6
♃	$\frac{6561}{8000}$	.820 125	.818 182	0.2
♂	$\frac{25}{36}$	.694 444	.690 048	0.6
⊕	$\frac{2916}{3125}$	.933 120	.930 669	0.3
♀	$\frac{243}{250}$	.972 000	.971 490	0.05
☿	$\frac{5}{12}$	.416 667	.427 083	2.4

These square roots are in inverse ratio as the apsidal distances corresponding to the motions in the first column, by Kepler's Area Law (to which he refers in the accompanying text, with a reference to Book V, Chapter III, part XII, where it is stated in the appropriate form). The next column contains the arithmetic mean of these two apsidal distances, that is the semi-major axis of the orbit (see figure 5.15). The next column contains the linear eccentricity, that is, a measure of the distance between the focus of the orbit and its centre, the numbers being found by subtracting the number in the previous column (i.e. a) from the larger of the two numbers in the column before that (i.e. $a + ae$). The linear eccentricity is then divided by the semi-major axis to give the true eccentricity of the ellipse, shown in the last column. Kepler explains the process of calculation in the accompanying text.

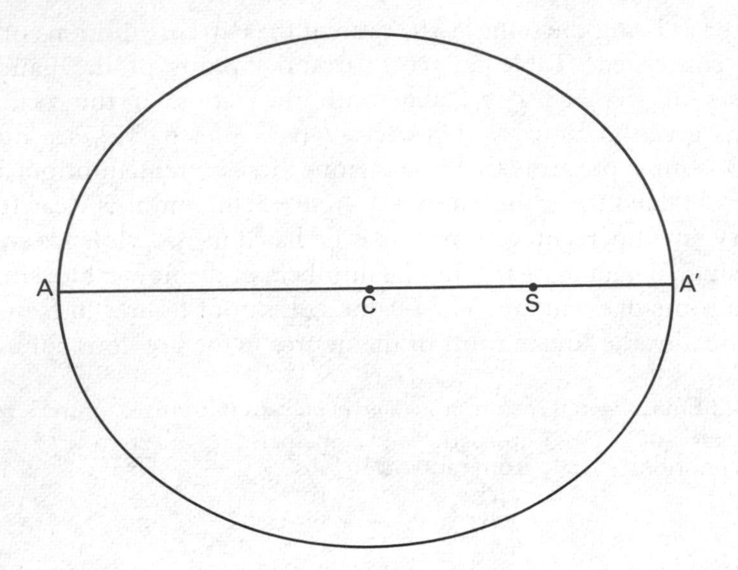

Figure 5.15 Elliptical Orbit

Aphelion A
Perihelion A′
Centre C
Focus S

Let semi-major axis, AC = CA′ = a, and eccentricity = e.
Then linear eccentricity, CS = ae, by definition.

$$\text{Aphelion distance, AS} = \text{AC} + \text{CS} = a + ae = a\,(1 + e).$$

$$\text{Perihelion distance, SA}' = \text{CA}' - \text{CS} = a - ae = a\,(1 - e).$$

$$\text{Major axis, AA}' = \text{AS} + \text{SA}'.$$

The next stage is to calculate the mean motions of the planets, whose relation to the extreme motions has been given in Book V, Chapter III, part XII. This calculation again involves the first two laws, and the results are again shown in a table (see figure 5.16). This table is rather more elaborate than the previous one since Kepler requires all the mean motions to be on the same scale, which necessitates finding a set of numbers which will represent the ratios of each of the motions to every one of the others. (The results in the previous table were pure numbers, so there were no problems of scale.) The

[P]roportio-nes Harmonicæ Binorum.	Numeri motuũ extremorum	singulorum ꝓpriæ.	Media singulorum continuata. Arithmeticum	Geometricum seu radix.	Differentiæ semisses.	Numerus motus medij in dimensione Sua	Communi.
1	♄.139968.	64	72.50.	72.00.	25.	71.75.	156917.
1	♄.177147.	81					
2	♃.354294.	6561	7280.5.	7244.9.	178.	72.271.	390263.
5	♃.432000.	8000					
24	♂.2073600.	25	30.50.	30.00.	25.	29.75.	2467584.
2	♂.2985984.	36					
32 3	T. .4478976.	2916	3020.500.	3018.692.	904	3017.788.	4635322.
5	T. .4800000.	3125					
5	♀.7464960.	243	246.500.	246.475.	125.	246.4625.	7571328.
1 3 8	♀.7680000.	250					
5	☿.12800000.	5	8.500.	7.746.	377.	7.369.	18864680.
4	☿.30720000.	12					

Figuræ ultra punctum pertinent ad præcisionem numeri in p. denarijs

Figure 5.16 Ratios of extreme motions in system of planets
(HM V, Ch. IX, p. 239)

mean motions, all on this arbitrary common scale, are shown in the last column. A note under the table draws attention to the fact that the numbers to the right of the points are decimal fractions. (Kepler usually uses sexagesimal fractions since he is dealing with angular measures most of the time.)

From the mean motions, Kepler calculates the mean dimensions of the planetary orbs, using his third law (given in Book V, Chapter III, part VIII). Since he has already found the eccentricities of the orbits, he can now calculate the radii of the aphelion and perihelion spheres of each of the planetary orbs (see figure 5.17) and the stage is accordingly set for a

	Numeri ex motibus medijs in dimensione Pristinâ,	Novâ eversâ inter cubos q̃rendâ.	Numeri ꝓportionis orbium inter quadratos inventi.	Semidiametri ut Supra.	Eccentricitas In dimensione Propriâ ut supra.	Communi	Intervalla extrema emergentia. Aphelia.	Perihelia.
♄.	156917.	29539960.	9556.	85.	5.	562	10118.	8994
♃.	390263.	11877400.	5206.	85222.	4222.	258	5464.	4948
T.•	2467584.	1878483.	1523.	55.	5.	138	1661.	1384
♂.	4635322.	1000000.	1000.	95178.	1647.	17	1017.	983
♀.	7571328.	612220.	721.	99295.	705.	5	726.	716
☿.	18864680.	245714.	392.	80625.	17375.	85	476.	308

Figure 5.17 Aphelion and perihelion distances (HM V, Ch. IX, p. 240)

Nam si Semidiameter orbis figuræ circumscripti quæ est communiter 100000, fiat	ex	Tunc Semidr inscripti pro	fiet	Cum sit intervallum ex Harmonijs.	
In Cubo	8994. ♄	57735	5194.	Medium	♃.5206.
In Tetraëdro	4948. ♃	33333	1649.	aphelium	♂.1661.
In Dodecaëdro	1384. ♂	79465	1100.	aphelium	T..1018.
In Icosiedro	983. T.	79465	781.	aphelium	♀. 726.
In Echino	1384 ♂	52573	728.	aphelium	♀. 726.
In Octaedro	716. ♀	57735	413.	medium	☿. 392.
In Quadrato Octaedri	716. ♀	70711	506.	aphelium	☿. 476.
vel	476. ☿	70711	336.	perihelium	☿. 308.

Figure 5.18 Dimensions of orbs, from polyhedral and harmonic archetypes (HM V, Ch. IX, p. 240).

comparison of the harmonic archetype with the modified version of the polyhedral one.

Kepler displays the comparison in one of his less clear tables (see figure 5.18). Taking the rubrics as the framework of a sentence into which the elements of the table are to be fitted, the sentence implied by the first row of the table would seem to be 'Now if the semidiameter of the circumsphere, which is usually taken as 100000, becomes 8994 instead, then the semidiameter of the inscribed sphere, for Saturn, instead of being 57735 [the value given in the table in Chapter XIV of the *Mysterium Cosmographicum*, see table 3.3 above] becomes 5194, while the distance found from harmonies is, for the mean sphere of Jupiter, 5206' (KGW *6*, p. 359, l. 10 ff). Table 5.3, which is a simplified version of Kepler's table, shows the polyhedron involved, then, in the second column, the assumed radius of the circumsphere, which is actually the radius of the perihelion sphere of the upper of the two planets in the fourth column. The third column shows the radius of the insphere corresponding to this circumsphere, the fourth column shows the planets involved (the insphere from the previous column refers to the second of these planets). The fifth column shows which sphere of which planet is being considered and gives its radius, found from harmonies. The last column shows percentage errors in the polyhedral values, taking the harmonic radii to be the 'true' values.

Comparing this table with the corresponding tables for the theory described in the *Mysterium Cosmographicum*, tables 4.4 and 4.5, suggests that Kepler modified the polyhedral archetype in order to bring it into better agreement with the observations, but only succeeded to a limited extent. The echinus gives a good fit to the aphelion sphere of Venus but

Table 5.3 *Dimensions derived from polyhedral and harmonic archetypes.* (Compare figure 5.18).

From polyhedra (poly.)				From harmony	harm.-poly. %
Polyhedron	circum.	in.	Planets	radii of spheres	harm.
cube	8994	5194	♄ to ♃	med ♃ 5206	−0.2
tetrahedron	4948	1649	♃ to ♂	aph ♂ 1661	−0.7
dodecahedron	1384	1100	♂ to ⊕	aph ⊕ 1018	+8.
icosahedron	983	781	⊕ to ♀	aph ♀ 726	+7.8
echinus	1384	728	♂ to ♀	aph ♀ 726	+0.3
octahedron	716	413	♀ to ☿	med ☿ 392	+5.4
square of oct.	716	506	♀ to ☿	aph ☿ 476	+6.3
or	476	336	☿ to ☿	per ☿ 308	+9.1

the orb of the Earth continues to give trouble, and Kepler emerges from his cosmological skirmishing with Mercury with much less honour than from his astronomical battle with Mars. He admits, in fact, to having been tempted to try to improve the agreement by increasing the eccentricity of the orbit of Venus and decreasing that of the orbit of Mercury. This gave better results for Mercury. 'But' he comments 'firstly I could not defend this decrease on harmonic grounds, for the new aphelion motion of Mercury would not fit into any Musical scale.' Moreover, one of his harmonic rules, stated earlier in the chapter as an axiom would no longer be valid, and 'finally, the daily mean motion of Mercury would become too large, and its period, which is the thing we know most accurately, would be made too short' (KGW *6*, p. 359, ll. 38–40, p. 360, ll. 1–4).

In his table, Kepler has compared the radii found from the polyhedral archetype with those found from the harmonic one. He does not go on to compare the radii found from the harmonic archetype with those calculated more directly from Tycho's observations. He probably thought that since his exact harmonic ratios were, as we have seen, very close to the 'observed' ratios of the velocities, then his perihelion and aphelion radii would be similarly close to those calculated more directly. In fact, this is not necessarily so – as is proved in any good lecture course on numerical analysis. A crude example is as follows: if we know two values to a precision of about three figures, and their true values are quite close, say 100.4 (corrected to 100.) and 100.6 (corrected to 101.), then if we take the difference of the values we know, that is 100. and 101., we get 1., but the actual answer should be only 0.2, and if we are then rash enough to use our answer later as a divisor, the result of the division will be too small by a factor 5, though our first two numbers were correct to 1%.

However, as it turns out, Kepler's 'harmonic' values are quite close to the values calculated more directly from observations, as can be seen from the first three columns of tables 5.4 and 5.5. The first column of each table shows the value found from Tycho's observations (see Bialas, 1971, and tables 4.4 and 4.5), the second shows the corresponding value calculated from the harmonic archetype, and the third column shows the error, as a percentage of the value in the first column. For interest, three further columns have been added, showing the modern calculated values for 1600 (see Bialas, 1971, and tables 4.4 and 4.5) and the errors in Kepler's two results expressed as a percentage of these 'true' values. It will be noted that in the cases marked with an asterisk, that is for the perihelia of Saturn, the Earth and Mercury and the aphelia of Saturn and the Earth the harmonic archetype gives more accurate values than those obtained by direct calculation. This is presumably due to the fact that the harmonic method makes use of the period of the planet, which is known with a precision unattainable for other elements of the orbits.

Table 5.4 *Perihelion distances*

	'observed'	harmonic	$\frac{\text{h-obs}}{\text{obs}}$ %	mod. 1600	$\frac{\text{obs-mod}}{\text{mod}}$ %	$\frac{\text{h-mod}}{\text{mod}}$ %
♄	8.9679	8.994	0.3	9.01193	−0.5	−0.2*
♃	4.9493	4.948	−0.03	4.95390	−0.09	−0.1
♂	1.3823	1.384	0.1	1.38195	0.03	0.1
⊕	.9820	.983	0.1	.98312	−0.1	−0.01*
♀	.71913	.716	−0.4	.71828	0.1	−0.3
☿	.30656	.308	0.5	.30753	−0.3	0.2*

Table 5.5 *Aphelion distances*

	'observed'	harmonic	$\frac{\text{h-obs}}{\text{obs}}$ %	mod. 1600	$\frac{\text{obs-mod}}{\text{mod}}$ %	$\frac{\text{h-mod}}{\text{mod}}$ %
♄	10.0521	10.118	0.7	10.09753	4.	0.2*
♃	5.4507	5.464	2.1	5.45170	−0.02	0.2
♂	1.6646	1.661	−0.2	1.66541	−0.05	−0.3
⊕	1.0180	1.017	−0.1	1.01688	0.1	0.01*
♀	.72915	.726	−0.4	.72838	0.1	−0.3
☿	.46955	.476	1.4	.46667	0.6	2.

It thus appears that in some cases Kepler's third law did, indirectly, give him more accurate dimensions for the orbs, though in the majority of cases the dimensions calculated from harmonies, with the help of the third law, are less accurate than those obtained more directly.

However, all the differences between Kepler's 'observed' and 'harmonic' values are small, being about an order of magnitude smaller than the errors that would be accepted by most modern cosmologists (in their own theories if not always in those of others). One has no need to suspend modern standards to accept that Kepler's conviction of the confirmation of his theory by observation was an entirely reasonable one.

Kepler and Ptolemy

The work with which *Harmonices Mundi Libri V* seems designed to be compared is, of course, *Timaeus*, though Kepler himself never explicitly makes this comparison, as far as I know. He would probably have felt it to be *lèse Platon* to do so, and he could, in any case, be sure that his readers would be at least reasonably familiar with Plato's work, which was available in several different editions by 1619. There is, however, another ancient work with which Kepler did compare his own, namely the *Harmonica* of Ptolemy.

At the time when Kepler became interested in Ptolemy's *Harmonica* the work had only been printed once, in a Latin translation by Antonio Gogava (Venice, 1562). Unfortunately, Gogava had worked from a rather corrupt Greek text and parts of his translation are quite seriously at fault as a result (see Düring, 1934). Kepler not unnaturally felt that he could do better than Gogava if he could obtain a Greek manuscript of Ptolemy's work. His, eventually successful, attempts to do so have been described in detail by Klein (1971). However, Kepler did not in fact produce the promised translation. The reasons for this are made fairly clear in *Harmonices Mundi* Book III and in the appendix to Book V which summarises Ptolemy's work and compares it with Kepler's own. It appears that Ptolemy's work no longer seemed very important to Kepler by 1618: its purely musical content had been developed entirely competently by Zarlino and others (as we have seen, Kepler continually refers to Zarlino's system as being Ptolemy's) and the application of the musical results to astronomy and astrology was indissolubly wedded to the Ptolemaic description of the planetary system and the astrological ideas of the *Tetrabiblos*.

Kepler's summary of Ptolemy's work, in an appendix to *Harmonices Mundi* Book V (KGW 6, pp. 369–73), is described as excellent by a modern editor and translator of the *Harmonica* (Düring, 1934). The explanation for its excellence is clearly the fact that in *Harmonices Mundi Libri V* Kepler had worked over very much the same ground as Ptolemy, though from a Copernican point of view and with different astrological beliefs. Kepler also differed from Ptolemy in seeking the origin of consonances in geometry rather than in the properties of numbers as such. All these differences are pointed out explicitly in Kepler's summary (KGW *6*, p. 372, l. 16 ff; p. 371, l. 37 ff; p. 370, l. 17 ff). However, Kepler's summary of the *Harmonica* equally makes it clear how close are the similarities between Ptolemy's work and his own. Indeed, there can be little question but that Kepler's work is essentially modelled upon Ptolemy's, and that in the *Harmonice Mundi* Kepler sees himself as rewriting Ptolemy's work in the light of modern knowledge – for the *musica mundana* described in both works has exactly the same scope. Like Ptolemy, Kepler sees musical consonances, the nature of astrologically powerful configurations and the structure of the planetary system as manifestations of a universal 'harmony' which is expressible in mathematical terms. For Ptolemy, the musical harmony is determined by the numbers which describe it – that is, it is numerological in origin – and the astrological harmony is an exact analogy of the musical one, involving the same ratios. For Kepler, as we have seen, the basis of the harmony lies not in numbers but in geometry, and the musical and astrological harmonies have different geometrical causes. This latter departure from Ptolemy seems to have been a consequence of Kepler's convincing himself, apparently on observational grounds, that there were not the same number of astrological Aspects as of musical consonances. Thus Ptolemy's simple analogy between the two sets of harmonies could no longer hold, and it is hardly unexpected that Kepler's earlier explanations, involving partial analogies and appeals to extra conditions, eventually gave way to separate explanations of the two sets of ratios. Nonetheless, as we have seen, Kepler does compare the two classifications arising from his two explanations. This appears to be a reminiscence of Ptolemy's scheme. Moreover, in pointing out the similarities in the two classifications, Kepler seems to be explaining the partial success of Ptolemy's theory in much the same way that in the *Astronomia Nova*, when rejecting an orbit of Mars compounded of circles in the manner of the *Almagest*, Kepler nevertheless points out that such an orbit was adequately exact for Ptolemy, since the planetary positions he used were only accurate to 10′ of arc (*Astronomia Nova*, Ch. XIX, KGW *3*, p. 177, l. 37 ff).

Once we recognise the close parallel between the *Harmonica* and the *Harmonice Mundi*, we can also see that Kepler's astrology is closely bound up with his cosmology. Astrological harmony is an integral part of his *musica mundana* as it was of Ptolemy's. Now, we know that Kepler eventually rejected almost everything of traditional astrology except the doctrine of Aspects (see Simon 1975, 1979 and Field, 1984c), so substantially all his astrological theory is closely bound up with his cosmology. This, it seems, accounts for the fact that Kepler appears to provide a counter-example to the otherwise suggestive simultaneity of the increase in respectability of Copernicanism and the decline in respectability of astrology (among the learned).

As Simon (1975, 1979) has shown, most of Kepler's reasons for rejecting standard astrological beliefs, such as the belief in 'houses', are closely connected with his Copernicanism. However, Kepler's belief in astrological *musica mundana* depends upon another element in his thought, one that classes him with the Old Guard rather than the *avant garde*, namely his belief that the Universe (or, at least, the observable Universe) is finite, and that the Sun, with its system of planets, holds a privileged position within it (see Koyré 1957 and Chapter II above). His cosmological models were thus neglected by later generations of cosmologists, who increasingly saw the Sun as but one star in an infinite Universe. So his metaphysical defence of astrology, being tied to an outmoded cosmology, exercised no influence. It thus appears that the example of Kepler, who was exceptional in that the consequences he drew from Copernicanism did not lead him to believe the Universe was infinite, does not in fact provide evidence against the theory that the rise of Copernicanism made a major contribution to the decline of astrology.

Since Kepler's *Harmonice Mundi* is so similar to Ptolemy's *Harmonica* one can hardly avoid the conclusion that Kepler's chief reason for not completing his translation of Ptolemy's work and printing it, as promised, as an appendix to his own, was his sense that the *Harmonica* was now only of historical interest. This is not, however, what Kepler says at the beginning of his appendix, where he apologises for the non-appearance of the translation promised on the initial title page, quoting a line and a half of Horace to the effect that 'the plan was to make an amphora but as the wheel turned a pitcher emerged instead' (KGW *6*, p. 369). The reasons Kepler gives here are that various difficulties arose and that his time was much taken up with other things.

The title page of *Harmonices Mundi Libri V* had promised that the appendix to Book V would compare Kepler's work not only with the *Harmonica* of Ptolemy but also with the 'harmonic speculations of Robert Fludd, found in his work on the Macrocosm and the Microcosm' (published in 1617 and 1618). Kepler in fact fulfilled the second part of his undertaking – quite briefly, and in a manner sufficiently lacking in controversialists' pyrotechnics for some modern historians to entertain the suggestion that Kepler himself was, in some sense, a Rosicrucian like Fludd. Despite the brevity of his treatment of Fludd's work, it seems that the matter of Kepler's relation to Fludd raises problems of a more general kind concerning Kepler's natural philosophy, and discussion of it will accordingly be postponed until we have drawn some general conclusions about Kepler's cosmological ideas, in the next chapter.

VI
Conclusions

The inspiration of the *Mysterium Cosmographicum* and of *Harmonices Mundi Libri V* is clearly Platonic, in the sense that in both works Kepler sets out to describe the beautiful mathematical Archetype according to which the observable Universe was constructed. Substituting the Christian God for Plato's gods and demiurges as the power behind the creative process appears to raise no problems, and, as we have seen, Kepler even goes so far as to state that he regards *Timaeus* as a commentary on *Genesis* (HM IV, Ch. I, KGW *6*, p. 221, see Chapter I above).

In a letter to Christopher Heydon, written in October 1605, Kepler was equally explicit about the source of the mathematical basis of the Archetype: 'Ptolemy had not realised that there was a creator of the world: so it was not for him to consider the world's archetype, which lies in Geometry and expressly in the work of Euclid, the thrice-greatest philosopher (*et nominatim in Euclide philosopho ter maximo*)' (Letter 357, lines 164–7, KGW *15*, p. 235).

The choice of mediator between *Timaeus* and the *Elements* is hardly less explicit: a long quotation from Proclus' *Commentary on the First Book of Euclid's Elements* appears in the original Greek on the title page of *Harmonices Mundi* Book I (see figure 5.1) and a Latin version of a slightly longer part of the same passage is to be found on the title page of Book IV. As we have seen in Chapter V, the import of this quotation is that the study of mathematical truths can lead to an understanding of the structure of the physical Universe.

Kepler and Proclus

As Caspar and others have remarked, Kepler felt particularly close to Proclus and makes frequent references to his works (KGW *6*, note to p. 13, p. 520). A reference in the introduction to *Harmonices Mundi* Book I (an introduction which seems to be intended as an introduction to the work as a whole, as explained in Chapter V) shows the extent of Kepler's respect for Proclus' Commentary on Euclid: 'In the four books he published on the first book of Euclid, Proclus Diadochus showed how a Theoretical Philosopher should treat mathematics' (KGW *6*, p. 15, ll. 16–17). Kepler then expresses his regret that Proclus has not left a commentary on the tenth book of the *Elements*.

Proclus' commentaries on Euclid and Plato, and a number of his original works on subjects as various as grammar and eclipses, were printed in Greek and in Latin translations during the sixteenth and early seventeenth centuries. It is quite possible that Kepler came across some of them when a student at Tübingen.

In astronomy, particularly in the matter of the significance to be accorded to observation, Proclus undoubtedly sees himself as a staunch supporter of Plato: his account of Ptolemy's *Hypotheses* even starts by recalling Plato's advice that the true philosopher should concentrate his attention upon an astronomy which deals in absolutes (Sambursky, 1962, p. 146). Sambursky (1962) shows Proclus as taking a rather harsh attitude to the astronomers' reliance on observations; Manitius' translations of the passages Sambursky quotes from the *Hypotyposis* are consistently rather less harsh; and Halma's are so much milder that Proclus' final, and unmistakable, condemnation of the astronomers' complicated mathematical 'fictions' seems slightly jarring. Nevertheless, there appears to be no reason to doubt that Proclus regarded the complicated mathematical models as just the kind of unrevealing formulation one would expect to obtain by unremitting devotion to observation. Sambursky (1962) seems to take this as a fair assessment of the state of geocentric astronomy at the time, but Kepler would surely have noted that the astronomers whom Proclus condemns include Aristarchus (Proclus *Hypotyposis* I1, trans. Sambursky, 1962, p. 146, passage 124). It seems that both in the *Hypotyposis* and in the short passages in the *Commentary on Timaeus* which relate to astronomy – passages to which Sambursky gives a harsher tone than that to be found in the translation by Festugière – Proclus' Platonic distaste for complicated models has compelled him to be rather dismissive about astronomical observation. Plato himself was in a happier position: the observations available in the fourth century B.C. were sufficiently crude to allow him to construct a very simple mathematical explanation of the motion of the Sun (*Timaeus*, 38c ff). Nevertheless, it is clear that this mathematical model is intended to account for the observations, and most modern scholars seem to agree that the passage about astronomical observation in the *Republic (Republic* VII, xi, 530c ff) should not be construed as advocating that astronomical theories should be constructed without recourse to observation (see, for example, Shorey's notes on this passage in the Loeb edition of the *Republic*, first printed in 1935). Since modern scholars have come to this conclusion, it seems quite reasonable to suppose that Kepler might also have interpreted Plato's work in a similar sense, and have felt that Proclus' most valuable contribution to astronomy was not his distrust of observation but his condemnation of the complexity of the mathematical fictions which the astronomers used to describe the motion of the planets, thus seemingly lending his support to the similar criticism which Copernicus was to make a thousand years later. Furthermore, it seems possible that it was from Proclus that Kepler derived the beginnings of his dislike of epicycles, a dislike we have seen expressed in the *Mysterium Cosmographicum* by his excluding epicycles from the planetary orbs (see Chapter III above and Grafton, 1973), and this at a time when, as Grafton (1973) has shown, he was still dependent upon Maestlin for help with some technical matters.

Despite their differences in regard to astronomy, it is nonetheless easy to understand why Kepler should have admired Proclus' work: Proclus shows a very firm grasp of mathematics as well as a deep interest in Plato. For example, the *Commentary on the First Book of Euclid's Elements* contains much that is of mathematical, as opposed to purely philosophical, interest, though the mathematical contributions are by no means always original to Proclus, as Heath (1921) and others have shown. Furthermore, Proclus was prepared to take his admiration of Euclid's rigour to the length of adopting Euclid's form of presentation, in recasting two books of Aristotle's *Physics* as series of propositions and proofs (Morrow, 1972, p. 161); and in his commentary on *Timaeus* he gives it as his opinion that Plato proceeds in the manner of the geometers, since before giving proofs he provides definitions and states hypotheses, and before embarking upon the study of Nature he enunciates the series of fundamental principles that will be his guide (Proclus trans. Festugière, 1966, vol. II, 236.15, pp. 66–7). This insight into the formal equivalence of mathematical and logical argument is crucial to the development of mathematical forms of logic, a highly significant branch of modern mathematics – though there is, of course, no need to suggest that Boole, Frege and their colleagues had been reading Proclus' commentary on *Timaeus*. Kepler, on the other hand, surely must have read it, though I have not succeeded in finding any reference to the work in any of Kepler's writings. Probably this silence (if it truly is silence) is to be explained by the fact that the surviving fragment of Proclus' commentary does not extend to the part of *Timaeus* which is explicitly mathematical, but breaks off after the description of the gods' creation of human souls (44e). Kepler no doubt regretted that Proclus' commentary did not include the mathematical part of *Timaeus*, but he was presumably spared an irritating certainty that falls to the lot of modern scholars: it has been shown that what survives is a fragment of a commentary that originally went right to the end of Plato's work (Proclus, trans. Festugière, 1966, vol. I, Preface, pp. 10–11).

Kepler's admiration for Euclid Trismegistus might well have sufficed without Proclus' example and encouragement, but it is notable that he shows a distinct preference for setting out his works in terms of definitions, axioms, propositions, etc., thus emphasising their formal relationship to mathematics proper. However, the fact that he occasionally committed the mathematical solecism of attempting to prove the truth of an axiom, for example in *Harmonices Mundi* Book IV (see Chapter V above), must be seen as an indication that in some cases he considered the relationship with geometers' reasoning to be rather superficial. His real purpose was, after all, that of the natural philosopher, as he had stated in the introduction to *Harmonices Mundi* Book I: '. . . I am not a Geometer working on [Natural] Philosophy, but a [Natural] Philosopher working on this part of Geometry' (KGW *6*, p. 20, ll. 1–2, see Chapter V above). Unlike the geometer, the philosopher is free to question his basic assumptions – as Proclus had remarked in his commentary on *Timaeus* (Proclus trans. Festugière, 1966, 236. 28–237. 3, vol. II, p. 67).

Historians have drawn another parallel between Kepler and Proclus, placing each at the end of the tradition he represents: Proclus as the last exponent of ancient Greek philosophy and Kepler as the last exponent of a form of mathematical cosmology that can be traced back to the shadowy figure of Pythagoras. The present study does not tend to confirm the validity of this portrayal of Kepler, but it seems likely that in one respect at least Kepler himself would have welcomed it: he too seems to have seen his cosmological work as definitive, if we may judge by the introduction to *Harmonices Mundi* Book V. This introduction begins with a discussion of the work of Ptolemy, including the *Harmonica*, before going on to describe

Kepler's discovery of his third law, which, as we have seen in Chapter V, had given him the tool he needed to show how the observed musical ratios between extreme velocities of the planets could determine the structure of the Solar system. His considerable exhilaration is thus understandable in the circumstances. The paragraph ends

> '. . . nothing holds me back, I can give myself up to the sacred frenzy, I can have the insolence to make a full confession to mortal men that I have stolen the golden vessels of the Egyptians to make from them a Tabernacle for my God far from the confines of the land of Egypt. If you forgive me I shall rejoice; if you are angry, I shall bear it; I am indeed throwing the die, and writing the book, either for my contemporaries or for posterity to read, it does not matter which: let the book await its reader for a hundred years; if God himself has waited six thousand years for his work to be seen.'[1]

Unfortunately, having led its army over the Rubicon and its people into the wilderness, Kepler's theory failed to emerge as a useful citizen of the promised land of modern science.

Kepler and Plato

Kepler's Platonism extends to his making use of mathematical arguments which are exactly parallel to the arguments used in *Timaeus* to establish the correspondences between the five elements and the five regular solids. We have seen that such arguments can be found both in the *Mysterium Cosmographicum* and in *Harmonices Mundi* Book V (see Chapters III and V above). However, though Kepler's arguments are closely similar to Plato's, the theories they are designed to establish are of very different characters. The theories of *Timaeus* seem to refer to the observable Universe rather obliquely, at the most indicating the mathematical lines along which a more detailed theory might be constructed (see Chapter I). There is no question of Plato's using his icosahedral forms of water to explain a measured value of the refractive index of ice. Kepler, on the other hand, does use his theories to explain quantitative properties of the observable Universe, properties such as the relative dimensions of the orbs of the planets.

One might, perhaps, expect the ancient influence to be seen most clearly in Kepler's earlier work, but it is in fact the later of the two cosmological treatises which is closer to *Timaeus*, both in its design and in its scope. Whereas the *Mysterium Cosmographicum* was a treatise on the heavens, conceived as the first of a series of cosmographic works, *Harmonices Mundi Libri V* was intended to deal with all those branches of natural philosophy which could be treated mathematically. Thus the earlier work starts in a 'modern' way, with an exposition of a limited problem, the spacing of the planetary orbs, which the author then proposes to solve, while the later more ambitious and less 'modern' work, being concerned with the cosmos as a whole, begins with mathematical fundamentals, involving two-dimensional figures as well as three-dimensional ones, and only later proceeds to show their application in the observed Universe.

In the *Mysterium Cosmographicum*, Kepler had decided that plane figures should be set aside, and an explanation of the spacing of the planets should be constructed by considering the properties of solid figures, since they shared the perfection of the Universe in having three dimensions and 'it is fitting that the Idea of the world should be perfect' (*Mysterium Cosmographicum*, Chapter II, KGW *1*, p. 25, ll. 15–16). When the second edition of the *Mysterium Cosmographicum* was published, this rejection of plane figures was marked by a note beginning 'Oh what a mistake' (KGW *8*, p. 50, l. 2, see Chapter IV above). The note went on to point out that plane figures must be considered, because God himself had employed plane figures in the Universe: the orbits of the planets are plane curves. Thus, in 1621, Kepler opposes a physical reason, that is a reason relating to an observed property of the Universe, to his earlier metaphysical preference for three-dimensional figures. However, he makes no attempt to answer the physical reason for using solid figures which he had put forward in the original preface to the *Mysterium Cosmographicum*, in the form of a question and answer: 'Why should there be plane figures between solid orbs? Solid bodies would be more appropriate' (KGW *1*, p. 13, ll. 5–6).

This concern to take account of the dimensionality of the problem under consideration seems to be characteristic of Kepler's thought. We find an example of it in *Harmonices Mundi* Book III, where Kepler states it more or less as a principle that the integers cannot explain musical consonances because (being dimensionless) they do not correspond to anything physical (see Chapter V above). In modern terms, this argument is fallacious, because the consonances are represented by ratios of lengths, and since both terms of the ratio have the same dimensionality the ratios themselves are dimensionless.

Kepler's line of reasoning suggests that in this passage, as in others (see Chapter V above), he is following the ancient Greek usage of regarding integers as numbers proper and treating rational numbers (that is, numbers expressible in the form a/b where a and b are integers) as if they were ratios. Such ideas might perhaps lead him to consider the dimensionality of the terms of the ratio rather than that of the ratio itself, making ratios of integers seem to have a different dimensionality from ratios of lengths. There is a happier example of Kepler's concern with dimensionality in *De Nive Sexangula* (Prague, 1611) where he insistently looks for a three-dimensional cause for the snowflake, since it comes into being in three-dimensional space. This contrasts sharply with Descartes' willingness to regard the hexagonal shapes as fragments of a flat sheet, their shapes being explained by the fact that regular hexagons will form a plane tessellation (Descartes, *Les Météores*, Leiden, 1637, Discours VI), an explanation Kepler explicitly excludes (*De Nive Sexangula*, p. 23, KGW *4*, p. 279, ll. 14–5). The contrast is made the more striking by the fact that whereas Kepler appears to believe that all snow crystals are flat, Descartes describes and illustrates columnar crystals of snow (see figure 6.1, diagram F).

Despite this concern with dimensionality, in *Harmonices Mundi Libri V* Kepler follows the example of *Timaeus* in using the properties

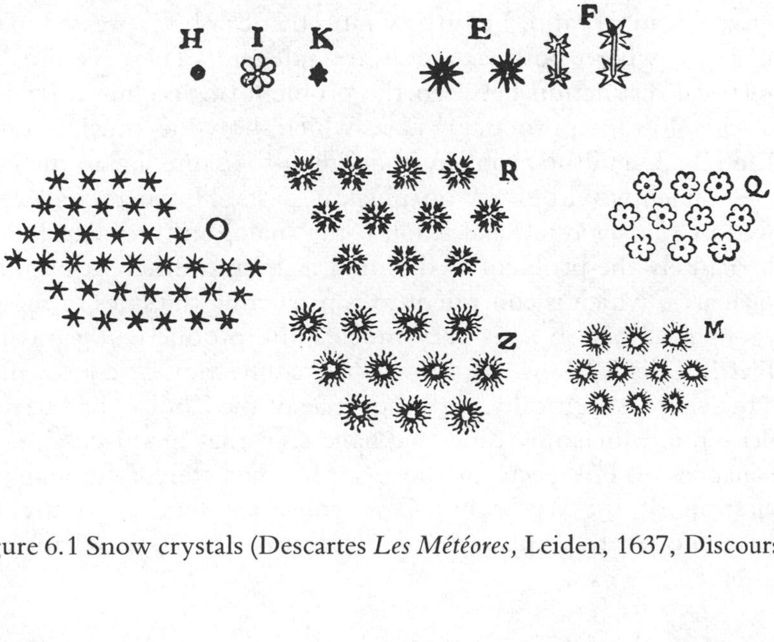

Figure 6.1 Snow crystals (Descartes *Les Météores,* Leiden, 1637, Discours VI)

of plane figures as a basis for his mathematical theory, though instead of Plato's triangles, which were introduced as being the simplest of polygons, Kepler considers the circle, the plane analogue of the simplest, and therefore noblest, of all mathematical figures, the sphere. The chosen figures are different, and Kepler goes on to consider not the properties of the circle itself but its relationship to figures derived from it and relationships among these figures themselves, but the principle embodied in the choice is the same: to start with the most fundamental mathematical figure.

The two-dimensional geometry of the first two books of *Harmonices Mundi Libri V* is, in fact, applied to two-dimensional problems: the generation of musical ratios among the arcs of circles and the generation of plane angles that correspond to astrological Aspects. As explained above, the problem of generating musical ratios is, in principle, dimensionless, but Kepler regarded it as involving lengths, and treated it as the problem of dividing a line in certain ratios. Now, it is a mathematical necessity that some curve, rather than a straight line, shall be the subject of the proposed division, since we require that there shall be a limit to the number of forms of division that may be obtained (there being a limit to the number of consonances), and there is no such limit in relation to the division of straight lines. For, by Proposition 9 of Book VI of the *Elements*, an nth part can be cut off from any straight line, where n is any integer (Euclid trans. Heath, 1956, vol. II, p. 211), and it can easily be shown that it follows that a straight line can be divided in any given rational ratio, i.e. in any ratio that can be expressed in the form $a : b$ where both a and b are integers. There is thus no geometrical distinction between the problem of dividing a straight line in the ratio 1 : 2, producing parts which show the 'musical' ratios 1 : 2 and 2 : 3, and the problem of dividing the same line in the ratio 17 : 19, which would give 'non-musical' parts. However, as Kepler showed in *Harmonices Mundi* Book I, the analogous problem for the circle, namely the problem of dividing it up into different numbers of equal arcs, which is equivalent to constructing the sides of regular polygons inscribed in it, is very much more productive of possible geometrical distinctions. Moreover, the connection of consonances with the circle, specifically with the circle of the Zodiac, had been in Kepler's mind for some time: we have seen that his discussions of consonances and Aspects in the years immediately following the publication of the *Mysterium Cosmographicum* tended to use the musical entities to explain the astrological ones (see Chapter V above). The idea that the arcs of the Zodiac corresponding to the angular distance between two bodies at Aspect were connected in some way with musical consonances is to be found in Ptolemy's *Harmonica* (Book III, Ch. 9) as Kepler remarked both indirectly in Chapter XII of the *Mysterium Cosmographicum* (KGW *1*, p. 42) and explicitly in his discussion of Ptolemy's work in the Appendix to *Harmonices Mundi* Book V (KGW *6*, p. 371, l. 38).

The three-dimensional problem of the spacing of the planetary spheres is treated very similarly in both Kepler's cosmological works. Since this problem is the main subject-matter of the *Mysterium Cosmographicum* and since, as we have seen, Kepler continued to believe that the solution proposed to it in that work was substantially correct, there was no need for him to go over all the same ground again in *Harmonices Mundi* Book V. His treatment of the matter in the later work is therefore quite brief; there is a rapid sketch of the theory in the comment following the proof that there are only five convex regular polyhedra, in *Harmonices Mundi* Book II section XXV, and a slightly fuller description of it appears in the summary of astronomical results in *Harmonices Mundi* Book V Chapter III. However, the theory is not actually applied in the simple form that is described in these passages. Instead, as we have seen in Chapter V, Kepler produces a modified form of it as a proposition near the end of *Harmonices Mundi* Book V Chapter IX. This modified form is not only slightly more complicated than the original one, but also lacks some of its power, since it does not always permit an exact calculation of the diameter of the aphelion sphere of a lower planet from the diameter of the perihelion sphere of the planet immediately above it. This weakening of the theory is not, however, crucial, for Kepler no longer has to rely upon the polyhedra as providing the Creator's sole Archetype: musical ratios among the velocities of the planets provide further criteria for making the Universe as beautiful as possible.

Kepler's final theory may seem to be over-defined, because, thanks to the third law, the musical ratios can be seen as determining the relative sizes of the outer and inner surfaces of the orbs of the system of planets (i.e. perihelion and aphelion distances) as well as the thickness of each orb (i.e. the eccentricity of the orbital ellipse – which can be found without using the third law). The apparent superseding of the polyhedral Archetype by the musical one is lent further colour by the fact that Kepler explains the appropriateness of each musical ratio to the particular pair of velocities between which it is observed. However, as we have seen in Chapter V, most of the proofs of the appropriateness of the musical ratios found among the extreme velocities of the planets depend, more or less directly, upon the geometrical properties of the Platonic solids associated with the spaces between particular pairs of neighbouring planets. As we have seen, Kepler regarded these associations as well-established: they were philosophically satisfactory and in reasonably good agreement with observation. He therefore felt able to derive part of his justification of the exact form of his musical Archetype from a consideration of the polyhedral Archetype. The remaining elements in the justification belong to music theory proper, but since, as we have seen, this too has been shown to be derived from geometrical results, the whole theory may be seen as entirely based upon geometry. The mathematical justification of the musical Archetype, in *Harmonices Mundi* Book V Chapter IX, is lengthier than the mathematical justification of the polyhedral Archetype, in Chapters III and VIII of the *Mysterium Cosmographicum*, because there are many more ratios between velocities than there are spaces between planetary orbs, and the style in which the material is presented resembles the *Elements* rather than *Timaeus*, but the purpose of the passages is clearly the same: to establish the mathematical coherence of the proposed Archetype, by explaining every element that might perhaps have appeared to be arbitrary.

There seems to be no logical reason why there should be only one most beautiful Archetype, but in *Harmonices Mundi Libri V* as in the *Mysterium Cosmographicum* Kepler is clearly determined that his Platonic Creator, committed as He is to creating the most beautiful of all possible Worlds, shall exercise no choice. This theological formulation seems the most natural one, since Kepler is concerned with an Archetype, but he is, of course, deducing the nature of this Archetype from a mathematical description of the present state of the observable Universe, and his tacit insistence upon the uniqueness of the Archetype is, in effect, equivalent to making the demand that there shall be a uniquely-determined mathematical cosmological theory which gives a coherent account of all the observations. In the first chapter of the *Mysterium Cosmographicum* he had stated explicitly that he believed such a demand could be fulfilled in relation to astronomical theories.[2]

Kepler's use of observations

Kepler explains in his preface to the *Mysterium Cosmographicum* that the observational evidence which suggested the polyhedral Archetype was the fact that there were six planets (in the Copernican description of the planetary system) (see Chapter III above). The musical Archetype described in *Harmonices Mundi* Book V has rather more complicated observational origins, being suggested (in principle if not in fact) by the 'observation' of musical ratios among extreme velocities of the planets. Since it is constructed from observed values of these ratios it cannot be tested against them in the way that the polyhedral Archetype was tested against the 'observed' radii of the planetary orbs. However, Kepler is at considerable pains to establish that he has done no violence to the observations in identifying the musical ratios from which the Archetype is constructed. Indeed, he is so scrupulous in noting where he has adjusted the 'observed' figures that at least one historian has supposed that the adjustment must have been significant (Koyré, 1961, p. 342) whereas it turns out to have been very small by modern standards (see Chapter V, table 5.1). It is not, however, clear what standards Kepler himself was applying.

It is well established that Kepler's concept of observational error was substantially the same as the modern one, in that it led him to reject a calculated orbit as inadequate because it failed to describe the position of the planet to within the margin of error claimed for the observed position.[3] Nevertheless, we should note that this particular form of comparison between 'theory' and 'observation' is a very simple one, since the theory in question (a suggested orbit) leads directly to a result which can be checked against observation (a planetary position). We may note also that, in this crucial passage in the *Astronomia Nova*, Kepler is making a false assumption about the behaviour of errors: there is no necessity for the error in the calculated position to be the same as the error in the observations that were used to make the calculation. Small errors can build up into large ones in the course of calculation, as in the crude example presented in Chapter V (p. 161). (This result is particularly important when the calculations are to be carried out by a machine, so that there is no intelligent supervision of the progress of the arithmetic – and it is accordingly proved in any university course on numerical analysis. At a more elementary level one finds the rule of thumb whereby the last figure of the answer is rounded off if mathematical tables have been used.)

The cosmological theories embodied in Kepler's polyhedral and musical Archetypes do not yield directly observable quantities in the straightforward way that the orbits in the *Astronomia Nova* yield them. Moreover, since the Archetypes refer to the overall structure of the Solar system they are essentially no more than descriptive – though there is no doubt that Kepler himself saw them as existing before the Universe was created (being mathematical Ideas and therefore coeternal with God) and thus as having the power to determine the, then future, structure of the material Universe. So for Kepler the Archetypes predicted this structure in the same way that a planetary orbit predicted positions of the planet. However, once the Universe has been created, the Archetypes are purely descriptive and can be checked against observation only in the sense of checking that they give an adequately accurate description of what is observed or deduced from observation.

This rather unsatisfactory descriptive character is also a feature of modern cosmological theories. Like Kepler's Archetypes they too deal with a very limited range of 'facts', which are at several removes from direct observation – for example, they are expected to 'predict' the expansion of the Universe and assign a value to Hubble's constant (the measure of the rate of expansion) whose 'observed' value seems to lie somewhere between 50 and 120 km per sec per kiloparsec (at the time of writing, i.e. Summer 1984). Kepler's Archetypes similarly deal in quantities which are not directly observable, namely the spacing of the planetary orbits and the ratios of extreme velocities of the planets.

The greatest contrast between Kepler and his modern counterparts lies in the degree of agreement they expect to find when the 'predictions' of their theories are compared with 'observed' values. Modern cosmologists regularly accept large errors, such as 50% or a factor of three either way, partly because they recognise that the numerical values they are attempting to explain are not very precisely determined, and partly because no modern cosmological model has ever produced more accurate results. Kepler, on the other hand, believed (correctly, as it turned out) that he was dealing with well-determined quantities; and his success in formulating astronomical laws which gave very accurate accounts of the observational data presumably encouraged him to believe that a cosmological theory might aspire to equal accuracy. However, assessing the degree of accuracy attained was no simple matter.

When he wrote the *Mysterium Cosmographicum* Kepler found

himself in a situation rather like that of his twentieth-century counterparts – and reacted to it in much the same way. He was aware that the radii of the planetary orbs were not very accurately determined and he noted, without undue emphasis, that his theory did not agree with them very closely in all cases. Tycho's observations modified this situation by allowing Kepler to calculate more accurate values for the radii of the planetary orbs. In fact, as Bialas (1971) has shown, Kepler's values for the major axes and eccentricities of the orbits are very accurate indeed (that is, they are in excellent agreement with the results of modern calculations of the values of these elements in 1600, see Chapter IV above), but Kepler himself had no way of knowing how much more accurate his newly-calculated dimensions were. Nevertheless, it is clear that since he believed his new dimensions to be considerably more accurate than those he had used in the *Mysterium Cosmographicum* he expected that they would be correspondingly closer to the theoretical values calculated from the polyhedral Archetype, if this had truly been the Archetype God had used in creating the Universe. Since it turned out that the agreement between theory and observation was not much changed by using the new more accurate orbits, Kepler set about modifying his theory.

The new, musical, Archetype shows clear affinities to the mathematical cosmology of Ptolemy's *Harmonica*, and to the geometrical Archetype described in *Timaeus*. It is less obvious, but surely at least equally significant, that the later, musical, Archetype shows an affinity to Kepler's astronomical theories, in fitting very closely with observational data. The temptation to cheat was certainly present – as we have seen in Kepler's comments on the orbit of Mercury at the very end of *Harmonices Mundi* Book V Chapter IX (see Chapter V above) – but Kepler seems always to have resisted it.

It is exactly this concern with the observable world that Kepler mentions as epitomising the difference between his own *Harmonices Mundi Libri V* and Robert Fludd's *Utriusque Cosmi . . . Historia* (Oppenheim, 1617 and 1618): 'So for him it is his conception of the World, and for me the World itself, or the real motions of the Planets in it, which are the subject of World Harmony' (HM V, App., KGW *6*, p. 377, ll. 1–3).

Kepler and Fludd

Kepler compares his work with Fludd's in the four final pages of the Appendix to *Harmonices Mundi* Book V. It should perhaps be noted that since the length of Kepler's work is about three hundred pages and the total length of the first two parts of Fludd's (with which the comparison is made) is about nine hundred and ninety pages, Kepler is contenting himself with a very brief analysis.

He begins by describing the structure of Fludd's work and indicating which passages of it relate to matters that are also considered in *Harmonices Mundi Libri V*. The plan of Fludd's work is fairly intricate, but Kepler's references are sufficiently detailed to allow a determined reader to locate the particular passages under discussion. They mainly concern 'artificial music', that is man-made music, which Fludd considers in his second book, *Utriusque Cosmi . . . Historia, Tractatus II* (Oppenheim, 1618), and 'world music', which is considered in the first book (Oppenheim, 1617). Fludd's ordering of these topics differs from Kepler's and seems to reflect an important divergence in outlook: whereas Kepler explains music theory in considerable detail (in *Harmonices Mundi* Book III) and then goes on to demonstrate the existence of a celestial counterpart to human music (in Book V), Fludd first gives a brief description of a very elementary form of heavenly harmony (doing little more than identify 'musical' ratios among the radii of various spheres in the Elemental, Celestial and Empyrean regions, see figure 6.2), and deals with music theory as a purely microcosmic phenomenon, going into considerable practical detail.

Kepler's criticism takes the two forms of music in his own order, starting with the 'artificial', which includes the theory of music. He first points out the differences in subject-matter, upon which his comments are neutral, but he also notes such differences as the fact that Fludd has accepted the ancient opinion that consonances are built up from smaller intervals, an opinion that he, Kepler, refuted in his own Book III, Chapter IV. He remarks that Fludd's work is unlike his own as 'a Practitioner differs from a Theoretician, for where he deals with instruments I inquire into the causes of things or consonances' (HM V, App., KGW *6*, p. 374, ll. 12–14). This comment appears to be just, and to touch upon an important point of difference between the works as a whole: Fludd's piece on 'artificial music' is a condensed version of a musical textbook, such as Zarlino's *Istitutioni Harmoniche* (Venice, 1558), whereas Kepler's piece, apart from being more purely theoretical (that is, consisting almost entirely of mathematics), is conceived as a section of an argument rather than as an article in an encyclopedia. As Kepler put it: 'In a word, in the study of Harmony, one of us shows himself a

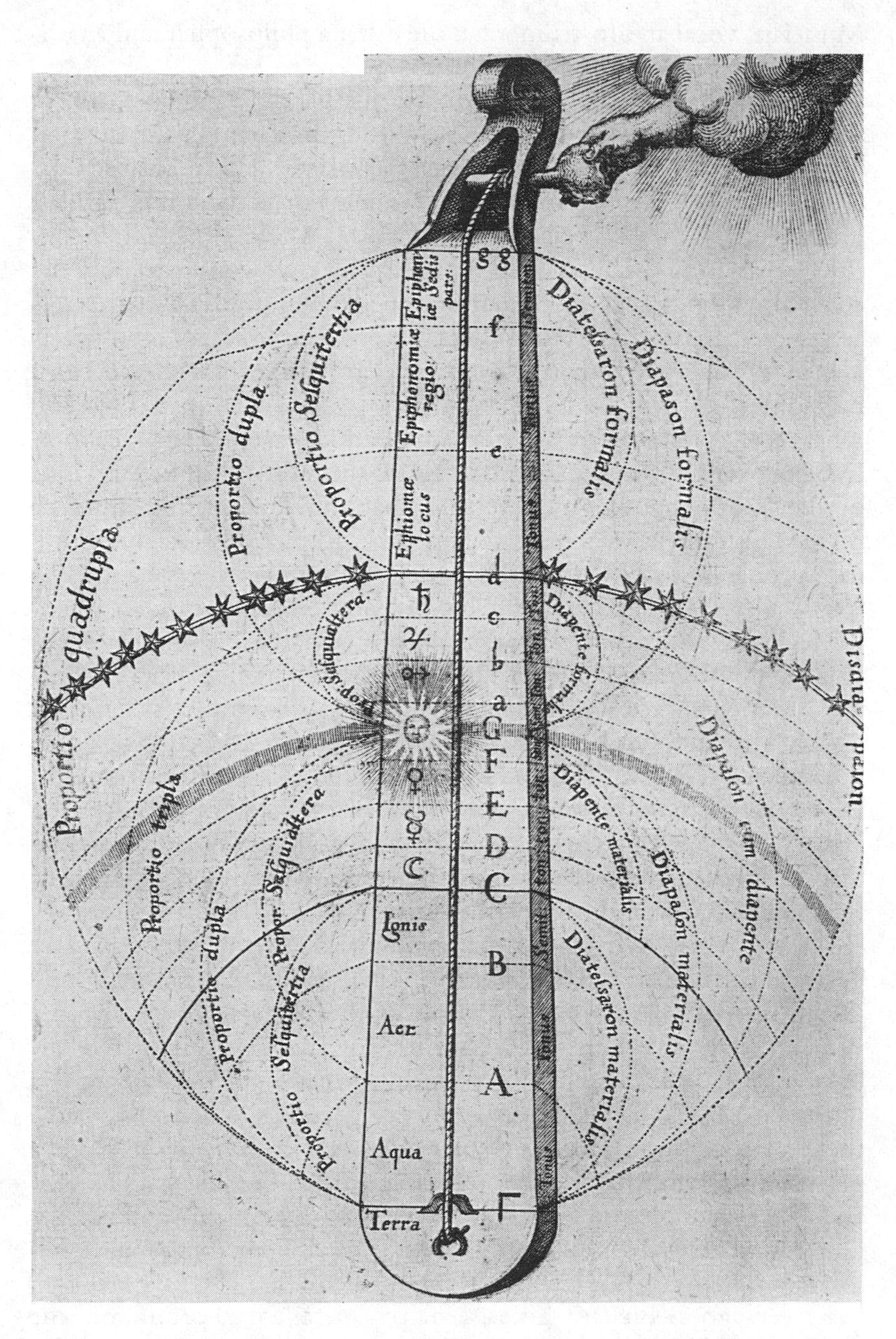

Figure 6.2 The Universal monochord (Fludd, *Utriusque Cosmi . . . historia*, Tractatus I, Oppenheim, 1617, p. 90)

Musician, vocal and Instrumental; the other a philosopher and Mathematician' (KGW *6*, p. 374, ll. 32–5).

After this summary Kepler turns to the part of Fludd's work 'in which he introduces Music into the World', prefacing his discussion with the statement that 'Here there is a vast difference between us' (KGW *6*, p. 374, l. 36). For example, Kepler correctly points out that Fludd's music concerns all three parts of the World – the Empyrean, the Celestial and the Elemental – whereas his own concerns 'only the celestial, and not even all of that but only the motions of the planets as it were beneath the Zodiac' (KGW *6*, p. 375, l. 24). Moreover, as Kepler says, Fludd adheres to the ancient belief that 'the force of harmonies is derived from pure numbers' (KGW *6*, p. 375, l. 26), whereas Kepler asks what the numbers measure. Furthermore, Kepler correctly notes that Fludd obtains the radii of his three regions of the World by merely asserting that the total radius must be divided into three equal parts, whereas he, Kepler, uses astronomical measurements to find the dimensions with which he is concerned (KGW *6*, p. 375, ll. 32–9). This use of assertion rather than an appeal to technical astronomy extends to Fludd's description of the planetary system, the only part of the World for which observational measurements were available. Fludd's planetary system is geocentric, and shows all the planetary spheres as being of equal thickness. This is, of course, conventional in expositions of Ptolemaic cosmology, and Copernicus follows the same convention in *De Revolutionibus* (see Chapter III above). However, Fludd seems disposed to take these equal dimensions quite seriously, for he marks them on his Universal Monochord (see figure 6.2), and it is necessary for his system of harmonies that the Sun shall lie exactly half-way between the two boundaries of the celestial region. Moreover, Fludd has retained the Ptolemaic ordering of the planets, which, by the time his diagrams were published, in 1617, had been rendered untenable (for those who chose to believe what Galileo saw through telescopes) by Galileo's observation that Venus showed phases like those of the Moon (1613). Kepler does not comment on Fludd's planetary system, possibly because he regarded the illustrations of it as further examples of what, in the section on man-made music, he had called Fludd's 'pictures', contrasting them with his own 'mathematical diagrams with letters on them'.[4]

Kepler goes on to give a fairly detailed account of the mathematical processes by which Fludd obtains his divisions of the World – for example by considering intersecting pyramids of light and darkness. The one quotation of Fludd's actual words is somewhat startling. Kepler says 'Nor does it diminish the diversity between us that he ascribes four degrees of obscurity and darkness to the Elemental region *because any thing*, he says, *has four quarters*, certainly no less so than three thirds, and five fifths: . . .' (KGW *6*, p. 376, ll. 19–22). Readers imbued with learnéd incredulity may be glad to know that the passage which begins on the eighth line of *Utriusque Cosmi . . . historia, Tractatus I*, Book III, Chapter II reads '*ideo necesse est, ut interiores terrae partes, cum quaelibet res ex quatuor quartis componatur, quatuor frigiditatis testimonia ob integram lucis absentiam retineant et possedeant*' (*Utriusque Cosmi . . . historia*, p. 82). Kepler has presumably quoted from memory since he makes a slight error in the words he quotes, but he has not deformed the sense: Fludd goes on to ascribe three-fold darkness to water, two-fold to air and only a single measure to fire. Moreover, the chapter from which this passage is taken seems to be fairly representative of Fludd's reasoning about the dimensions of the World: he is concerned with ascribing dimensions to regions whose extent cannot be measured by observation and whose very existence was open to dispute, namely the spheres of water, air and fire which he takes to lie between the body of the Earth and the sphere of the Moon (see figure 6.2). I am happy to leave it to others to decide what Fludd's purpose may have been, but it seems clear that he is not concerned with the observed properties of the planetary system, for he never (apparently) makes any allusion to observed values of any astronomical quantity. Kepler's judgement that Fludd is concerned with a World of his own imagining may be over-simple, but there seems to be no reason to doubt that it was a World very different from the one with which Kepler himself was concerned.

Kepler's motive in embarking on this critique of Fludd's work seems to have been that since it was, like his own, concerned with 'harmonies' he felt it would be appropriate to point out where the works agreed and where they differed (HM V, App., KGW *6*, p. 373, l. 13 ff). As we have seen, for the most part they differed. For the most part, also, Kepler's tone is polite, though his lack of sympathy with Fludd's work is tolerably obvious, and his distaste occasionally shows through in such remarks as the complaint that Fludd's Hermetic analogies 'are dragged in by the hair' (HM V, App., p. 375, ll. 2–6). Fludd's retort was to be that Kepler heaped up definitions, axioms and propositions (Fludd *Demonstratio*, Frankfurt, 1621, Text XV, p. 25). Apart from their non-reportorial tone, both comments appear to be entirely justified – and revealing of the very different preferences of the authors as to what constitutes a convincing style of argument.

However, such illuminating exchanges are not typical of the brief controversy that ensued upon Kepler's critique of *Utriusque Cosmi . . . historia*. Fludd's reply to Kepler's three sides (folio) was 54 pages (folio): *Veritatis proscenium . . . seu Demonstratio quadam analytica . . . in appendice quadam a Joanne Keplero, nuper in fine Harmoniae suae Mundanae edita . . .* (Frankfurt, 1621). He criticises Kepler's summary of the few sections of *Utriusque Cosmi . . . historia* that were in question, but most of his book is concerned to point out where he and Kepler disagree, for example about Copernicanism, he, Fludd, is in the right. It is, moreover, clear that Fludd has been unable (or unwilling) to follow much of Kepler's mathematical argument and has largely construed it as similar in style to his own.

In view of Fludd's very strange account of the *Harmonice Mundi*, it is not surprising that Kepler wrote a reply (50 folio pages): *Pro suo opere Harmonices Mundi apologia. Adversus Demonstrationem Cl.V.D. Roberti de Fluctibus Medici Oxoniensis* (Frankfurt, 1622) (KGW *6*, pp. 361–457). He corrects Fludd's misrepresentations of the *Harmonice Mundi*, but by way of answer to Fludd's criticism of Copernicanism the reader is mainly referred to Kepler's other works (almost *all* Kepler's other works).

Fludd replied to this with a work of 83 quarto pages: *Monochordum Mundi symphoniacum seu Replicatio Roberti Flud . . . ad apologiam . . .* (Frankfurt, 1622). Most of this last work (pp. 19–75) is taken up with a refutation of Copernicanism, in counter-refutation of Kepler's reply to Fludd's comment on one particular part of the Appendix (Text XII in Fludd's *Demonstratio*). In fact, throughout their exchanges, Kepler and Fludd mainly confine themselves to discussing, in order, the particular passages of Kepler's Appendix that Fludd cited in his *Demonstratio* (actually he cited nearly all of it, dividing it up into very brief sections). Thus the succession of rival opinions can be read as a series of individual dialogues concerning particular texts. Since in Fludd's *Replicatio* his side of the dialogues was almost entirely concerned with rejecting Copernicanism, it is hardly surprising that Kepler, who had already referred Fludd to his astronomical works, did not feel it necessary to make any further reply.

As we have seen, one of the contrasts Kepler points out between his own work and that of Fludd is that Fludd's harmonies ignore actual units and use abstract numerical relationships, whereas Kepler finds musical ratios among quantities measured in the same units, such as the extreme angular speeds of planets as seen from the Sun (HM V, App., KGW *6*, p. 375, l. 25 ff). Fludd picks this up as Text XVII, and launches into a defence of numerology:

> 'In this passage I see that the author is entirely ignorant of the true numbers of natural Harmony: . . . Yet he describes Pythagoras' triangular number [the Tetractys, 1 to 4] on page 4 of his Book III[5] . . . He tries to avoid abstract numbers in his harmony; yet it seems that without using abstract numbers nothing can genuinely (*sincere*) be expressed in numbers, for no less abstract are the Mathematical numbers from lines, surfaces and bodies, or roots, squares and cubes, than are those found in common Algorithmic Arithmetic. The wiser philosophers, Themistius, Boëthius, Averroes, Pythagoras and Plato . . .' (Fludd *Demonstratio*, XVII, p. 25, l. 13 to p. 26, l. 5).

According to Fludd, numerological explanation is applicable not only to music but even at the highest level:

> 'Further, all kinds of natural things, and those which are supernatural, are bound together by particular formal numbers. The mystery of these occult numbers is best known to those who are most versed in this science, who attribute the Monad or unity to God the artificer, the Dyad or duality to Aqueous Matter, and then the Triad to the Form or light and soul of the universe, which they call virgin.' (Fludd *Demonstratio* XVII, p. 26, l. 17 ff)

The numerological creed which Fludd advances here seems to be in perfect accord with what we find in *Utrisque Cosmi . . . historia*. It has been quoted in preference only for its greater concision. Fludd's numerology is, however, less radical than Kepler's commitment to geometry: he is prepared to use geometrical methods in the same way that he uses arithmetical ones. For example, he explains the fact that the Sun's orbit is mid-way between the two boundaries of the celestial region by appealing to two intersecting pyramids (as he calls them), one of light radiating from an aureoled triangle that represents the Trinity, and another of darkness whose base lies in a plane through the centre of the body of the Earth (see figure 6.3).[6] However, the monochord which defines the more detailed structure of the celestial region is purely arithmetical (see figure 6.2).

The difference between Fludd's cosmological mathematics and Kepler's at least partly reflects a difference in their technical capacity as mathematicians, the advantage all being on Kepler's side.

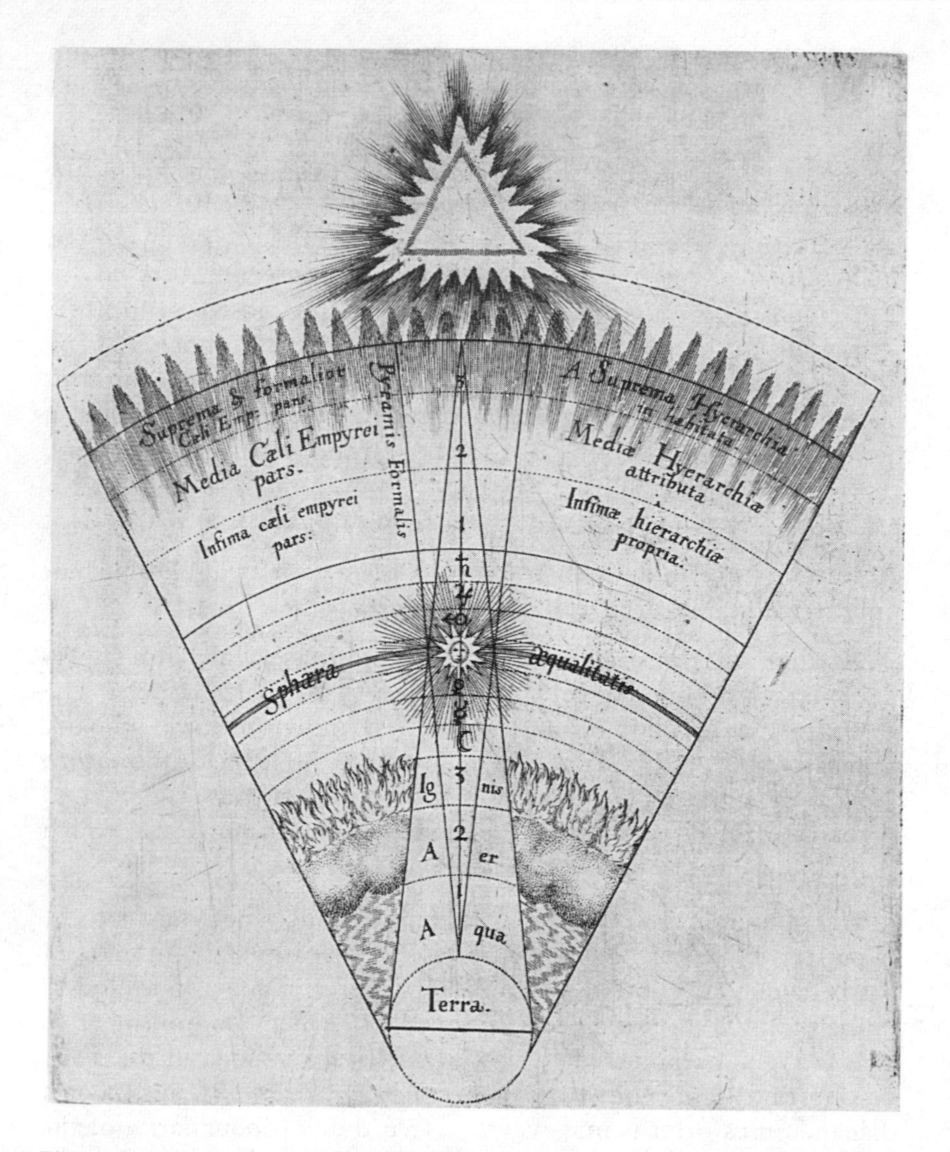

Figure 6.3 Intersecting pyramids of light and darkness, from Fludd *Utriusque cosmi . . . historia*, Tractatus I, Oppenheim, 1617, p. 89.

However, there does not seem to be a huge philosophical difference between Kepler's theory that the spacing of the planetary orbs was taken from the radii of the circumspheres and inspheres of the five Platonic solids and Fludd's theory that the mean radius of the orb of the Sun was determined by the geometrical properties of a pair of intersecting pyramids. One can hardly help reflecting that the latter theory is a somewhat elaborate way of constructing a mid-point, but it is nonetheless clear that a similarly Platonic God is geometrizing in both theories.

Fludd is more conventional than Kepler in being willing to accept numerology as well as geometry as a possible source of cosmological theories, and here it seems we have found a difference in their philosophies rather than merely a difference in their technical capacity as mathematicians: for Kepler, as we have seen, rejects numerology not on the grounds of its mathematical naïveté but on the grounds that numbers can have no properties apart from the quantities they measure. Nevertheless, it seems that the crucial difference between Kepler's World Harmony and Fludd's does not lie in the nature of the mathematics that is used, and it is quite understandable that historians who have concentrated their attention upon the thoroughgoing Platonism of Kepler's mathematical cosmology have concluded that Kepler has much in common with Fludd. The crucial difference between Kepler and Fludd seems rather to be that Kepler demanded that his cosmological theories should be in good numerical agreement with measured properties of the observable Universe.

Kepler's mathematical philosophy

It is clear that Kepler felt that a sharp distinction should be drawn between the types of natural philosophy exemplified by Fludd's *Utriusque Cosmi . . . historia* and his own *Harmonices Mundi Libri V*. Nevertheless, historians are not bound to agree with him, and the recent revival of interest in possible connections between the magical Hermetic tradition and the rise of modern science has led to several suggestions that Kepler, who is undoubtedly implicated in the latter phenomenon, may have had some connection with the former (to which Fludd clearly belongs).

Westman (1977) and others have considered problems such as whether Hermetic philosophers were truly drawn to the Copernican theory as an astronomical theory or whether they in fact saw it more or less as a symbolic representation of the power of the Sun. However, it seems that no-one has yet taken up the specific suggestion that Yates' *Rosicrucian Enlightenment* 'has provided historical material from which a new historical approach to Kepler might be made' (Yates, 1972, p. 223). The present study of Kepler's geometrical cosmology does not claim to be an adequate response to this challenge, but it has, at least, led us to make a fairly detailed analysis of *Harmonices Mundi Libri V*, a work which Kepler himself considered capable of comparison with what he recognised as an overtly Hermetic work, namely *Utriusque Cosmi . . . historia*. It seems reasonable to include the *Mysterium Cosmographicum* in the comparison as well, since, as we have seen, the theory it describes reappears as one element in the wider theory described in the later work, and the use of mathematics in the two works is similar in spirit (if not always identical in detail). Most historians would certainly agree that both of these works may fairly be said to 'look Hermetic' in a way that, for instance, the *Dioptrice* (Augsburg, 1611) clearly does not.

There is a further reason for looking to Kepler's cosmological works for traces of Hermetic philosophy: it is in writing about cosmology, rather than about the technical details of natural philosophy, that a man's philosophical prejudices and preoccupations are liable to be at their most visible. Cosmologists are synthesising the results of technical astronomy and the 'observations' they are dealing with are thus even more heavily theory-laden than those of the astronomers, with the result that their hypotheses tend to take the form of 'principles', such as the Cosmological Principle (that the Universe looks the same to all observers) or Mach's Principle (that the gravitational field at any point is determined by all the matter in the Universe).

The philosophical outlook in Kepler's cosmological works seems to be best described by some phrase such as 'radical Platonism'. We have seen that Kepler regards his thought as being closely akin to that of Proclus, but the numerous references to his works all seem to concern mathematical and philosophical points which show the Platonic rather than the Neoplatonic elements in Proclus' philosophy. The adjective 'radical' seems to be warranted by the fact that Kepler (who is rather given to laying his opinions on the line) states in so many words that *Timaeus* should be read as a commentary on *Genesis* (see Chapter I above). Kepler himself occasionally confuses the issue by describing his opinions as Pythagorean, or, rather, by ascribing his own opinions to the Pythagoreans – for example, in *Harmonices Mundi* Book II section XXV (KGW *6*, p. 81), where he suggests that the Pythagoreans may have used the five 'cosmic' solids in the manner described in the *Mysterium Cosmographicum*. However, it is entirely clear, from passages such as his introduction to his musical theory in *Harmonices Mundi* Book III, that he does not share the characteristically

Pythagorean belief in the power of pure numbers, even in the form in which it is expressed in *Timaeus* (34b–36d, see Chapter V above).

Another Pythagorean attitude which Kepler does not share is their belief that results should be kept secret or revealed only to initiates. Kepler's conviction appears to have been quite the reverse, if we may judge by the quantity of his publications and the many passages in his letters that echo the words he wrote to Maestlin in 1595: 'If this [i.e. the *Mysterium Cosmographicum*] is published, others will perhaps make discoveries I might have reserved for myself. But we are all ephemeral creatures (and none more so than I). I strive, therefore, for the glory of God, who wants to be recognised from the book of Nature, that these things may be published as quickly as possible. The more others build on my work the happier I shall be' (Kepler to Maestlin, 3 Oct. 1595, letter 23, l. 251 ff, KGW 13, pp. 39–40). This shows an entirely 'modern' attitude to publishing one's results, and it is in perfect accord with the impression one receives from Kepler's books, namely that they were written to be understood by the learned community as a whole – unlike, for instance, the beautifully illustrated and thoroughly enigmatic works by Michael Maier which Yates regards as typical products of Rosicrucianism (Yates, 1972, p. 70 ff). Of course, a work may be partly written for initiates without being as ostentatiously opaque to outsiders as, say, Maier's *Atalanta Fugiens* (Oppenheim, 1617): *Timaeus* may well be such a work, as Kepler seems to have been aware. Moreover, it is impossible to prove that any writer was not a covert adherent of a more or less secret group of adepts: one is forced to rely upon a subjective assessment of the writer's character. My assessment of Kepler's is that he would not have hesitated to tell his readers if his works contained an element of Pythagorean secrecy or if he was in some passages writing for initiates in the manner of Plato's *Timaeus*. Such admissions would have done no more than define further details of his professed allegiance to the philosophies he recognised as most clearly expressed in his works. For a historian they might have served as an indication that he shared some beliefs that may be regarded as typically Rosicrucian.

One is similarly forced to argue mainly from silence in assessing the likelihood that Kepler was influenced by the Hermetic tradition. We have seen that the references to Hermes in *Harmonices Mundi Libri V* are evidence only that Kepler has read *Pimander*, and we have seen also that he follows his summary of the work by stating that he disagrees with much of what it says. The only reference to Bruno in

Harmonices Mundi Libri V concerns his edition of Euclid (HM I, sect. XLVIII, KGW *6*, l. 12). It seems that the only conclusion that one can draw from Kepler's silence is that, since he does seem to have been in the habit of acknowledging his intellectual debts, as may be seen, for instance, from his many references to Proclus, Copernicus and Tycho, his silence on Hermes and his followers implies that he did not see himself as influenced by a tradition they represented. The character of the natural philosophy in his cosmological works inclines one to suppose that he was right about this: *Harmonices Mundi Libri V* shares with Hermetic and Rosicrucian works only those elements that can be seen as most directly derived from *Timaeus*. Fludd may have taken his Platonism from Hermes, Kepler's seems to have come from the primary source.

In any case, we need not be at a loss for a tradition to which Kepler may be assigned: the tradition is that of the hard-working professional mathematicians who were the spiritual children of Claudius Ptolemy. It is undoubtedly to this tradition that Kepler's astronomical work belongs. Moreover, we have seen that Kepler's musical Archetype is a development of the musical scheme described by Ptolemy in his *Harmonica*. Kepler thus has a double allegiance to his Alexandrian predecessor, for his own cosmological Archetypes arose from the possibility of turning the long Ptolemaic tradition of astronomical calculation to the task of solving a cosmological problem, namely explaining the overall structure of the planetary system: in the Copernican description of the planetary system the relative sizes of epicycle and deferent in the geocentric orbit (which could be calculated from observations) allowed one to go on to calculate the physical size of the planet's heliocentric orbit (in terms of the radius of the Great Orb). In keeping with the standards of this Ptolemaic tradition, Kepler demanded that his cosmological theories, like his astronomical ones, should be in good numerical agreement with the results of astronomical observation. Kepler's historical judgement is notoriously unreliable but there is, after all, a certain appropriateness in the references to measurement in the terse epitaph he wrote for himself:

> I used to measure the Heavens, now I measure the shadows of Earth.
> The mind belonged to Heaven, the body's shadow lies here.
>
> (KGW *19*, p. 393).

The Platonism of his cosmological theories makes them seem far removed from the 'modern' values that are apparent in, say, the *Astronomia Nova*, but Kepler's concern with giving an adequately accurate description of measured properties of the observable Universe appears to run through all his work.

Appendix 1
Diameters of stars

Kepler left this as an exercise for the reader in Chapter XXI of *De Stella Nova* (Prague, 1606, KGW *1*, p. 253, l. 40). See Chapter II for details. Let the Earth be at T and the centres of the stars at S_1 and S_2. The stars are given to be equidistant from the Earth, so let $TS_1 = TS_2 = d$.

Let the angle subtended by each star at T be α.

Then $\angle MTS_2 = \frac{1}{2}\alpha$

$\therefore$ radius of star,

$$S_2M = d \sin \tfrac{1}{2}\alpha = NS_2$$

Let the angle between the stars on the sky be θ.

Then $S_1S_2 = 2d \sin \frac{1}{2}\theta$.

Let the angle the second star subtends at the first be β.

$$\text{Then } \sin \tfrac{1}{2}\beta = \frac{NS_2}{S_1S_2} = \frac{d \sin \frac{1}{2}\alpha}{2d \sin \frac{1}{2}\theta} = \frac{\sin \frac{1}{2}\alpha}{2 \sin \frac{1}{2}\theta}$$

Note that this is independent of d.

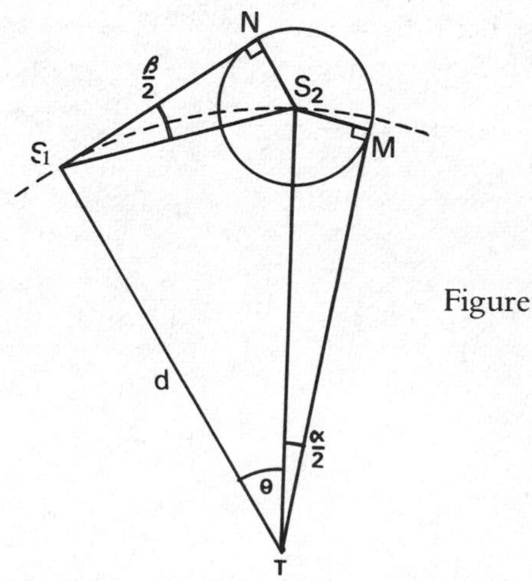

Figure A1.1

Since we are dealing with very small values of α and θ we may use the approximations

$\sin \frac{1}{2}\alpha = \frac{1}{2}\alpha$ and $\sin \frac{1}{2}\theta = \frac{1}{2}\theta$, where the angles are measured in radians.

$\therefore \sin\frac{1}{2}\beta = \frac{1}{2}\cdot\frac{\alpha}{\theta}$

Substituting the given values $\alpha = 2'$ and $\theta = 81'$ we obtain

$\sin \frac{1}{2}\beta = \frac{1}{2}\cdot \frac{2}{81}$ (since the conversion factor to radians cancels out)

$= 0.0123456 \ldots$

$\therefore \frac{1}{2}\beta = 43'$, roughly (using 4-figure tables)

$\therefore \frac{1}{2}\beta = 86'$

$= 1°26'$

This figure agrees well with the result of 'nearly three solar diameters' Kepler obtained in the *Epitome*.

Dropping the assumption that stars of equal brightness are at equal distances from the Earth

Although the assumption that stars of equal apparent brightness are at equal distances from the Earth is not an unreasonable one, Kepler seems to have felt uneasy with it, for he proceeds to show that his argument is still valid if this assumption is relaxed (*De Stella Nova*, Ch. XXI, KGW *1*, p. 254 f, trans in Koyré 1957, p. 66).

Let us therefore consider two stars which are of the same apparent brightness, that is, they subtend equal angles at the eye of the observer on the Earth, but which are at unequal distances. This case is shown in figure A1.2, where, as before, the centres of the stars are S_1, S_2 and each subtends an angle α at the Earth, T, while the more distant star, centre S_2, subtends an angle β at S_1.

Kepler asserts that β is always greater than α.

Let the radius of the star centre S_2 be r and the distances of the stars from the Earth d_1, d_2. Let $S_1S_2 = x$.

As before, from $\triangle NS_1S_2$ we have

$$\sin\tfrac{1}{2}\beta = \frac{NS_1}{S_1S_2}$$

$$= \frac{r}{x} \qquad (1)$$

From $\triangle MTS_2$

$$\sin \tfrac{1}{2}\alpha = \frac{r}{d_2}$$

therefore

$$r = d_2 \sin \tfrac{1}{2}\alpha$$

Substituting this value for r in (1) we have

$$\sin \tfrac{1}{2}\beta = \frac{d_2}{x} \sin \tfrac{1}{2}\alpha \qquad (2)$$

We have assumed that S_1, S_2 are close on the sky. Therefore $\angle S_1TS_2$ is very small. Therefore
$\angle TS_1S_2 + \angle TS_2S_1 \cong 180°$.

If $d_1 = d_2$ then $\phi_1 = \phi_2 \cong 90°$, but if d_2 is considerably greater than d_1 then ϕ_1 will be obtuse and we shall have

$$d_2 > x.$$

(For if K is the foot of the perpendicular from T to the line S_1S_2 then it is clear that, since ϕ_1 is obtuse, $x < S_2K$. But $S_2K < d_2$, because d_2, being the hypotenuse of the right-angled triangle TS_2K, must be its longest side.

$$\text{Therefore} \quad x < S_2K < d_2,$$
$$\text{therefore} \quad x < d_2.)$$

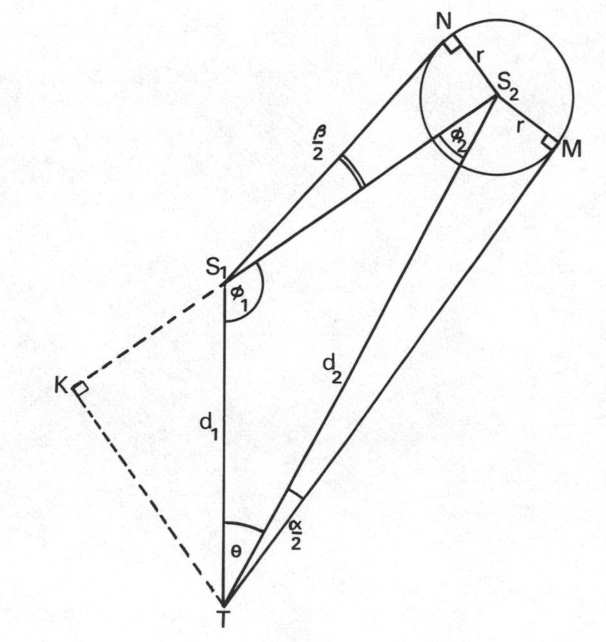

Figure A1.2

We have seen above that

$$\sin \tfrac{1}{2}\beta = \frac{d_2}{x} \sin \tfrac{1}{2}\alpha \qquad (2)$$

Now, since $\qquad d_2 > x,$

we have $\qquad \sin \tfrac{1}{2}\beta > \sin \tfrac{1}{2}\alpha$

therefore $\qquad \beta > \alpha.$

That is, the second star, seen from the first, appears brighter than either star appears from the Earth.

I am grateful to Dr A. E. L. Davis for suggesting the main lines of the proof given in the second part of this appendix.

Appendix 2
Radii of the circumsphere and insphere of the cube

Chapter XIII of the *Mysterium Cosmographicum* (KGW *1*, p. 43) gives a fairly cursory account of how to construct circumspheres and inspheres for the five known regular polyhedra. In this account Kepler refers particularly to Book XV of Campanus' Latin edition of Euclid's *Elements* (first printed in Venice, 1482, many later editions). The other edition to which he refers is that by François Foix de Candale (Paris, 1566 and 1578).

There was no need for Kepler to give a detailed account of how to construct the spheres, since the problems were thoroughly familiar to any mathematician of the time, having been treated *in extenso* by such popular authors as Luca Pacioli (whose work on polyhedra was very deeply indebted to a treatise written by his teacher Piero della Francesca (*c.* 1416–92)). The following modern account of how to calculate the radii of the spheres for the cube is intended merely to show that the problem is not a very difficult one.

Let ABCDEFGH be a cube, with centre O.

Let the centre of the face ABCD be P.

Let the side of the cube be 2*a*.

Since O is the common centre of the circumsphere and the insphere the radii of the spheres are clearly OP and OC respectively.

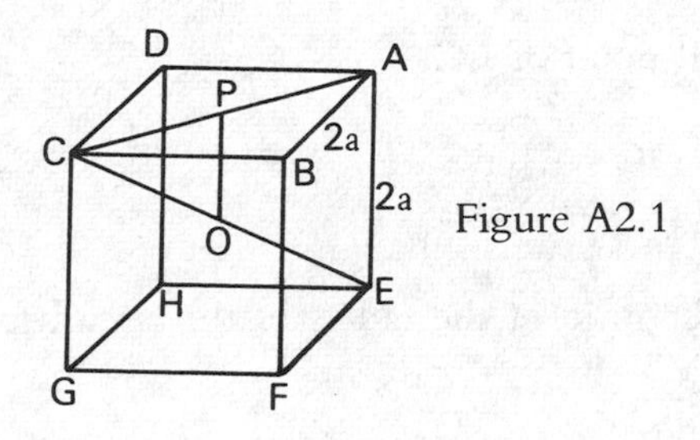

Figure A2.1

Since P and O are the mid-points of CA and CE respectively, it is clear that the triangles CPO and CAE are similar, their corresponding sides being in the ratio 1 : 2.

Therefore PO $= \frac{1}{2}.AE$
$= a.$

In $\triangle ABC$, by Pythagoras' Theorem we have

$$CA^2 = AB^2 + CB^2$$
$$= 8a^2.$$

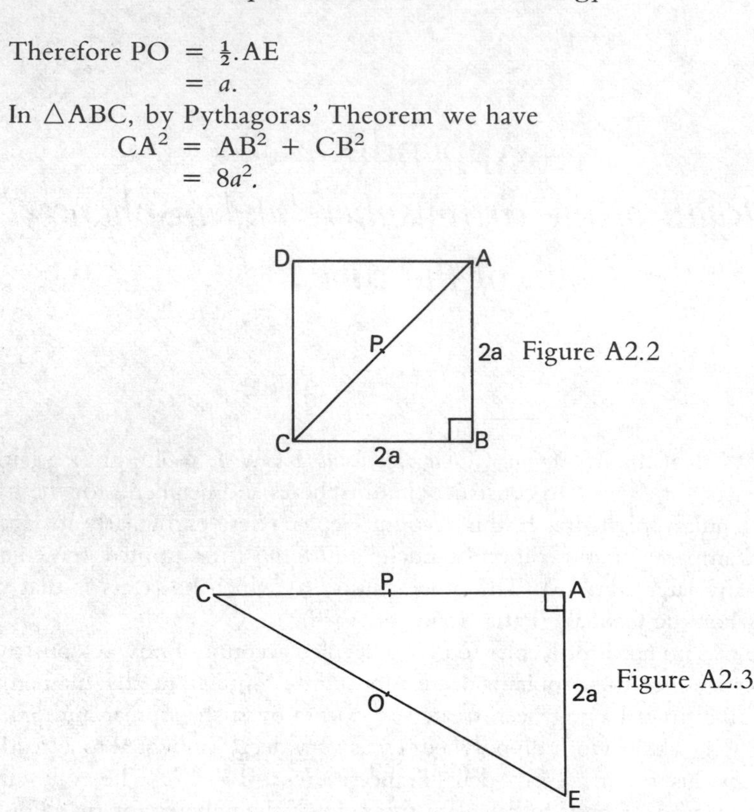

Figure A2.2

Figure A2.3

In $\triangle ACE$, $\angle EAC$, is a right angle, since AE, being a side of the cube, is perpendicular to the face ABCD and therefore to the line AC.

Therefore, by Pythagoras' Theorem,

$$EC^2 = AE^2 + CA^2$$
$$= 4a^2 + 8a^2$$
$$= 12a^2.$$
$$EC = 2\sqrt{3}a.$$

O is the mid-point of EC.
Therefore

$$OC = \frac{1}{2}.EC$$
$$= \sqrt{3}a.$$

Therefore the ratio of the radius of the circumsphere to the radius of the insphere is $\sqrt{3} : 1$.

Appendix 3
Radii of circumcircles and incircles of regular polygons

Problems of inscribing and circumscribing circles in or about polygons are dealt with in Book IV of Euclid's *Elements*. References in the *Mysterium Cosmographicum* (Tübingen, 1596) suggest that while Kepler was writing this work he was, in the main, using the edition of the *Elements* by François Foix de Candale (Paris, 1566).

Euclid shows how to inscribe and circumscribe circles in or about triangles, squares, regular pentagons and regular hexagons. He also shows how to inscribe any of these figures in a given circle and how to circumscribe it about a given circle. The final proposition, Proposition 16, is the problem of inscribing a regular fifteen-sided figure in a given circle.

In the main text of the *Mysterium Cosmographicum* Kepler treats the inscribing of regular polygons in circles as irrelevant to cosmological problems, but he was to change his mind later, and Book I of the *Harmonice Mundi* (Linz, 1619) contains a detailed analysis of the general problem of inscribing a regular polygon in a circle, the treatment being derived from both Book IV and Book X of the *Elements*. An account of this part of Kepler's work is given in Chapter V above. The present Appendix is concerned only with the theory described in the Preface to the *Mysterium Cosmographicum*.

The equilateral triangle

In Book IV of the *Elements* Euclid considers a general triangle, but Kepler is concerned only with an equilateral triangle.

Let PQR be an equilateral triangle. By *Elements* Book IV, Propositions 4 and 5, it is possible to construct circles inscribed in the triangle and circumscribed about it, as shown. Let the centre of the circumcircle, PQR, be O.

Then, by symmetry, O is also the centre of the inscribed circle, and PO produced will meet QR at L, the point in which QR touches the inscribed circle. Let the side of $\triangle$PQR be $2a$ and the radii of the inscribed and circumscribed circles b and c respectively.

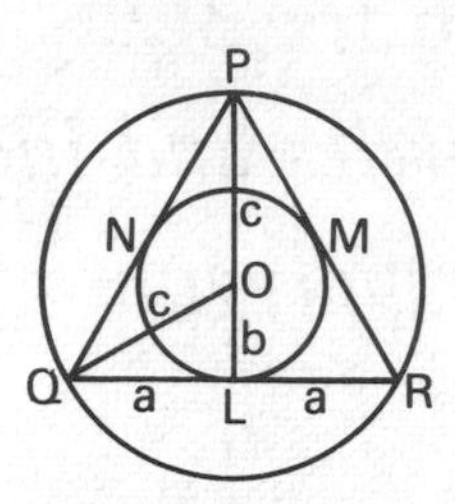

Figure A3.1 Equilateral triangle

$$\text{Then} \qquad \text{Area } \triangle PQR = \tfrac{1}{2}.QR.PL$$
$$= \tfrac{1}{2}.2a.(b+c)$$
$$= a(b + c). \tag{1}$$

But we may consider $\triangle PQR$ as made up of six triangles congruent with $\triangle OQL$.

$$\text{Therefore, Area } \triangle PQR = 6 \times \text{Area } \triangle OQL$$
$$= 6 \cdot \tfrac{1}{2} \cdot QL \cdot OL$$
$$= 3.a.b. \tag{2}$$

Therefore, by (1),

$$3ab = ab + ac$$
$$c = 2b.$$

So the ratio of the radii of the circumcircle and the incircle is 2 : 1. (The ratio of the radius of the inner surface of the sphere of Saturn to the radius of the outer surface of the sphere of Jupiter, deduced from the *Prutenic Tables* (Tübingen, 1551), is about 1.69 : 1. The agreement with the ratio found from the triangle is not startlingly exact. See *Mysterium Cosmographicum* Preface, KGW *1*, p. 12.)

The square

Let PQRS be a square. By *Elements* Book IV, Propositions 8 and 9, it is possible to construct circles inscribed in the square and circumscribed about it, as shown. By symmetry, the two circles are concentric. Let their common centre be O. Let the side of the square be 2*a*.

$\triangle PKO$ is an isosceles right-angled triangle.

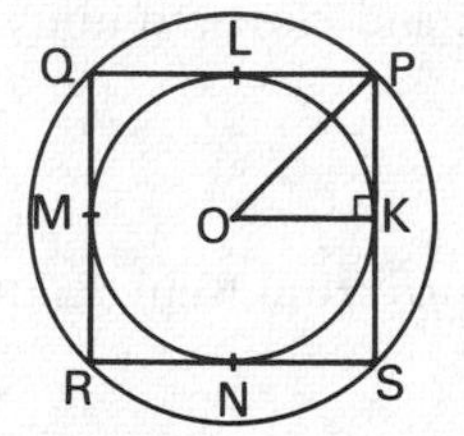

Figure A3.2 Square

$$\therefore \text{OK} = \text{PK} = a,$$

and, by Pythagoras' Theorem,

$$\text{OP}^2 = 2.\text{OK}^2$$
$$\text{OP} = \sqrt{2}\, a.$$

The ratio of the radii of the circumcircle and the incircle is thus $\sqrt{2} : 1$, that is, approximately 1.4 : 1. (The ratio of the radius of the inner surface of the sphere of Jupiter to the radius of the outer surface of the sphere of Mars, deduced from the *Prutenic Tables*, is about 2.98 : 1. The ratio from the square is thus too small by a factor of about two. See *Mysterium Cosmographicum*, Preface, KGW *1*, p. 12.)

Higher regular polygons

In general, it is clear, from considerations of symmetry, that it is possible to construct a circumcircle and an incircle for any regular polygon, and that the two circles will be concentric.

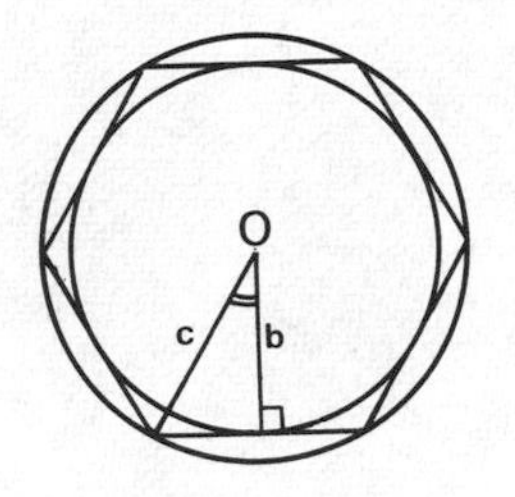

Figure A3.3 General case

Since the angle subtended at the centre by half of one side of an n-sided polygon will be $(360/2n)°$, the ratio of the radii of the circumcircle and incircle will be $\sec(360/2n)°$ as can be seen from figure A3.3. As n gets larger, this ratio gets closer and closer to unity.

Conclusions

The numerical values obtained from Kepler's theory are summarised in the table.

Table A3.1 *Orbs and polygons*

Planet	ratio of radii of spheres	Polygon	ratio of radii of circles
Saturn			
	1.69	triangle	2
Jupiter			
	2.98	square	1.41
Mars			
	1.35	pentagon	1.23
Earth			
	1.35	hexagon	1.15
Venus			
	1.54	heptagon	1.11
Mercury			

The best agreement is for the space between the Earth and Mars, but even that value is accurate only to about 10%. It is perhaps an indication of the low degree of numerical accuracy that Kepler was prepared to tolerate in a theory of this kind that he does not mention these very large errors as a reason for rejecting the theory – though it should be noted that it is not clear that the two objections he does raise are in fact intended to refer to the naïve form of the theory shown in the above table. (See *Mysterium Cosmographicum*, Preface, KGW *1*, p. 12.)

Appendix 4
Kepler and the rhombic solids

The rhombic dodecahedron and the rhombic triacontahedron are described and illustrated in *Harmonices Mundi* Book II (Linz, 1619) see figure 5.3 (diagrams Vu and Xx) and figure A4.1. Though these are Kepler's first published illustrations of the solids, the dodecahedron had been described, and the triacontahedron merely mentioned, in *De Nive Sexangula* (Prague, 1611, p. 7, KGW *4*, p. 266, ll. 10–16). Both solids reappear in Book IV of the *Epitome* (Linz, 1620) where we find a description of the structure of the dodecahedron (*Epitome* IV, Pt. I, p. 461, KGW 7, p. 270) and illustrations of both solids (*Epitome* IV, p. 464, KGW 7, p. 272). Together with the cube, which Kepler, entirely reasonably, regards as being in principle also a solid with rhombic faces, the rhombic dodecahedron and triacontahedron are later pressed into service to explain the fact that Jupiter has four moons and to account for the sizes of the gaps between their orbits, though Kepler does not describe his theory in detail (*Epitome* IV, Pt. II, p. 554, KGW 7, p. 318, l. 39 ff).

Of these published references to the solids, the one in Part I of *Epitome* Book IV is the most satisfactory from the mathematical point of view. As we noted

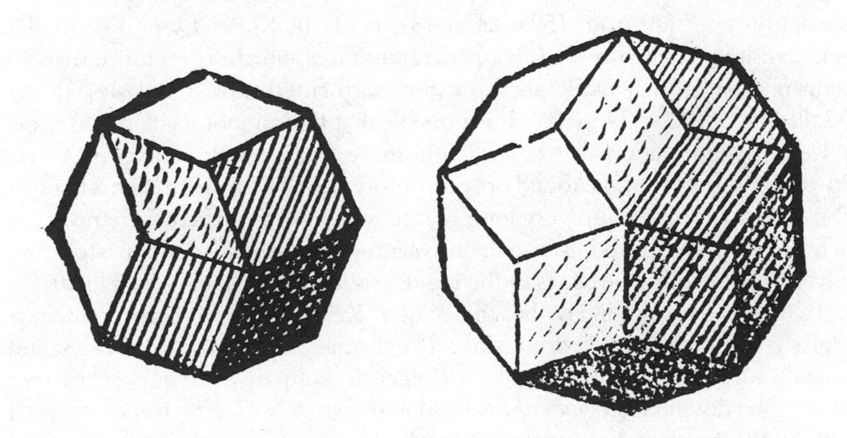

Figure A4.1 Rhombic dodecahedron and rhombic triacontahedron, from *Harmonices Mundi* Book II, p. 61.

in Chapter V, the proposition in *Harmonices Mundi* Book II has been phrased in such a way as to avoid rigorous mathematical discussion; and, though the rhombic dodecahedron is described in some detail in *De Nive Sexangula*, the rhombic triacontahedron is merely mentioned as another solid of a similar kind.

Kepler's description of the rhombic dodecahedron in *De Nive Sexangula* is very thorough, so thorough as to suggest that he knew of no earlier description of the solid, and regarded himself as in some sense its discoverer. One can hardly avoid the reflection that the discovery of the rhombic dodecahedron should have followed closely upon the first detailed inspection of a well-constructed honeycomb. However, Kepler does not refer to any earlier work, and since he does apparently make a habit of acknowledging the work of his predecessors, and there seems to be no reason why he should have departed from his usual practice in this case, it seems we must conclude that he regarded his investigation of the properties of the solid as largely original. The rhombic triacontahedron does not present itself to view in the natural world in any similar manner, and historians seem to be agreed in accepting that Kepler discovered this solid, though there is a discreet silence on the subject of the dodecahedron (see Caspar's notes on HM II, sect. XXVII, KGW *6*, p. 532 and on letter 142, KGW *14*, p. 467).

The date of Kepler's discovery

Kepler mentions both rhombic solids in a letter to Maestlin written in November 1599: 'Meanwhile I rely upon getting the eccentricities from the rhombic solids with 12 and 30 faces and from the Archimedean solids' (22 Nov. 1599, letter 142, ll. 21–2, KGW *14*, p. 87). A line in a letter to Maestlin written in the previous August also seems to refer to at least one rhombic solid: '. . . and there are the five regular solids and the 13 Archimedeans. And perhaps some Keplerian solids (*Et forsan aliquot Kepleriana*), of one of which I shall give you a description . . .' (29 Aug. 1599, letter 132, l. 141 ff, KGW *14*, p. 46). Kepler then describes a solid with twelve pentagram faces which 'is no more than an augmented [Platonic] Dodecahedron (but augmented most regularly)' (letter 132, ll. 144–5, KGW *14*, p. 46). It is possible that the 'aliquot Kepleriana' refers to Kepler's expectation that he will find more bodies with pentagram faces. However, in the event, he found only one more such body, and it seems likely it would not have taken him very long to satisfy himself that there were no more such solids to be found, since their construction depended upon stellating polygons formed by the edges of the regular solids. (For details see Field 1979a, Pt. I). It thus seems unlikely that the 'aliquot Kepleriana' are meant to include Kepler's second star polyhedron, since, like the first, it too has twelve faces, and it would surely have been natural for Kepler to compare and contrast the two solids (as he does in *Harmonices Mundi* Book II sect. XXVI) if he had discovered both of them when he wrote this letter. He would clearly have expected Maestlin to be interested, as is indicated by his giving a quite detailed description of the first star polyhedron. It thus seems that since 'aliquot' must refer to at least three bodies, the bodies in question were the 'augmented Dodecahedron' and the two rhombic solids.

The manner of Kepler's discovery

Kepler has not left an account of how he discovered the rhombic solids, but it seems probable that the description of the rhombic dodecahedron in the *Epitome* (Book IV Pt. II), which forms an extension to a mathematical discussion that is repeated from a passage in the *Mysterium Cosmographicum*, indicates the actual process of discovery of the rhombic triacontahedron. It certainly indicates a possible process of discovery. The references to the rhombic solids in the unfinished treatise on geometry which Kepler began to write in 1628 seem to be too sketchy to give any guidance (see Field 1979a; the treatise is printed in KOF VIII. 1, pp. 174 ff).

In Chapter V of the *Mysterium Cosmographicum*, Kepler mentions a relation between the three 'primary' regular solids, the cube, the tetrahedron and the dodecahedron: '. . . the remaining four [regular solids] are not generated by their faces but are either cut out from the cube, like the tetrahedron, by the removal of four right-angled pyramids, or are built up from it, like the dodecahedron, by the addition of six five-faced solids' (KGW *1*, p. 31, ll. 17–19). What Kepler says of the tetrahedron and the dodecahedron is true (see figures A4.2 and A4.3), and the octahedron may also be thought of as 'cut out' from a cube (see figure A4.4), but his statement is not quite true of the icosahedron, which requires the cube to be 'built up' into a dodecahedron and the icosahedron to be 'cut out' from the dodecahedron (see figure A4.5). There are no

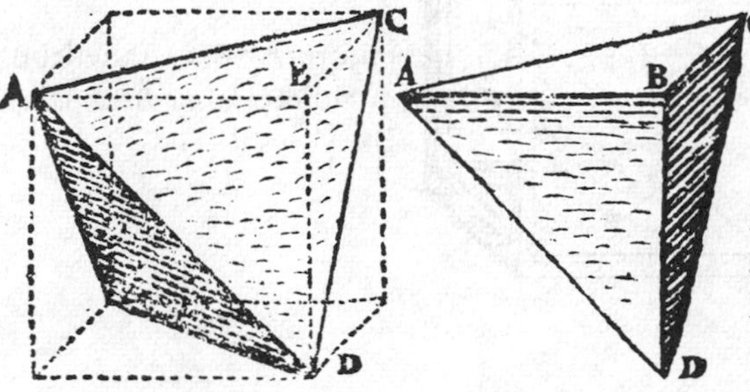

Figure A4.2 Tetrahedron and cube (*Harmonices Mundi* Bk V, p.181)

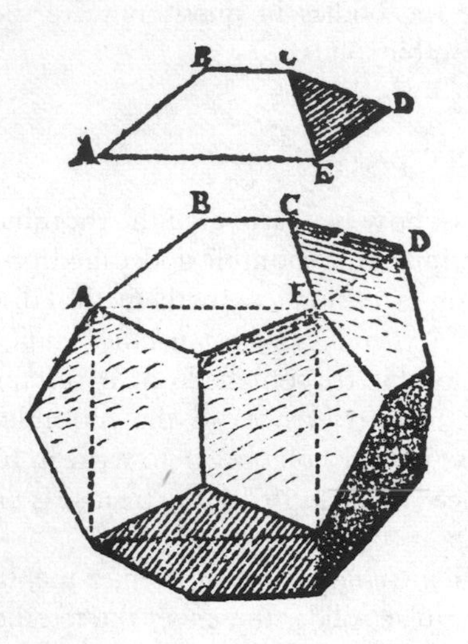

Figure A4.3 Cube and Platonic dodecahedron (*Harmonices Mundi* Bk V, p. 181)

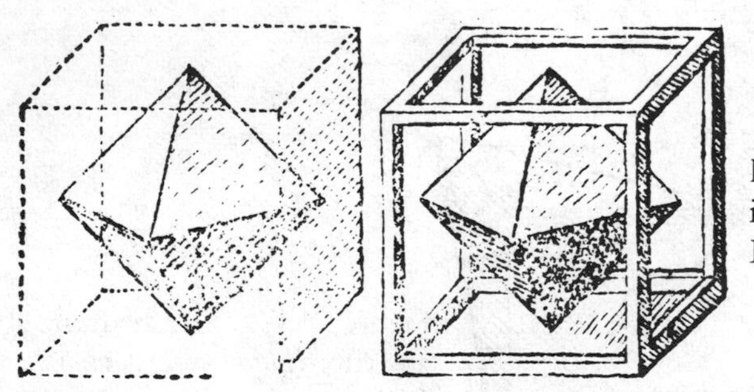

Figure A4.4 Octahedron in cube (*Harmonices Mundi* Bk V, p. 181)

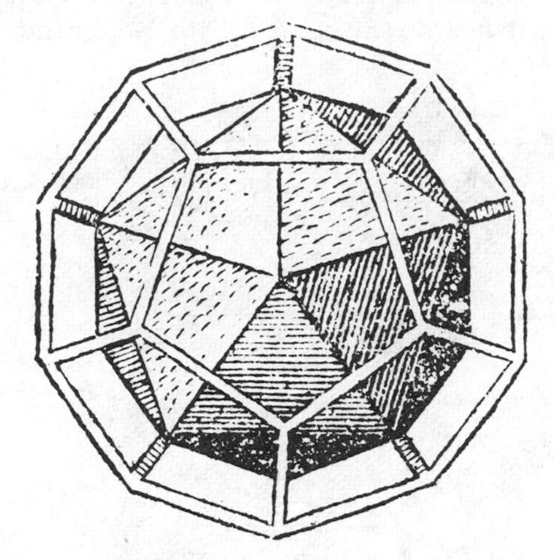

Figure A4.5 Icosahedron in Platonic dodecahedron (*Harmonices Mundi* Bk V, p. 181)

illustrations to this passage of the *Mysterium Cosmographicum*. Figures A4.2, A4.3, A4.4 and A4.5 are taken from illustrations to the corresponding passages in *Harmonices Mundi* Book V (Linz, 1619).

The idea of cutting pieces off a mathematical solid or adding pieces to it was not a new one in Kepler's day. Pacioli describes the results of such processes in *De Divina Proportione* (Venice, 1509) and the names of the processes, 'cutting' and 'raising', are reflected in the titles given to the resulting solids in the spectacular diagrams supplied by Leonardo da Vinci (see figures A4.6 and A4.7). It should be noted, however, that the pyramids Pacioli adds to the faces of the solids are always equilateral. The same processes are employed by Wentzel Jamnitzer in his *Perspectiva Corporum Regularium* (Nuremberg, 1568), a book which consists almost entirely of illustrations, designed by Jamnitzer himself and engraved by his chief engraver Jost Amman (see figures A4.8 and A4.9). Jamnitzer occasionally adds pyramids which, though not themselves regular or equilateral, will produce a regular solid. Such a case is shown in figure A4.8, where the pyramids added in the middle figure on the right seem to have made an exact cube, but not an exact octahedron. In the absence of any explanatory text, it is impossible to decide what Jamnitzer thinks he is doing. However, one strongly suspects he is primarily concerned with some mathematical version of Alchemy, with religious overtones, rather than with the mathematical properties of polyhedra or the process of drawing in perspective.

Kepler's ulterior motives are cosmological, but he is explicit about the geometrical methods he uses. He is concerned that the resultant solid should exhibit some form of regularity, rather than that the pieces added should themselves be regular or equilateral. Thus, whereas Pacioli has added equilateral pyramids to the tetrahedron (see figure A4.10), Kepler adds pyramids

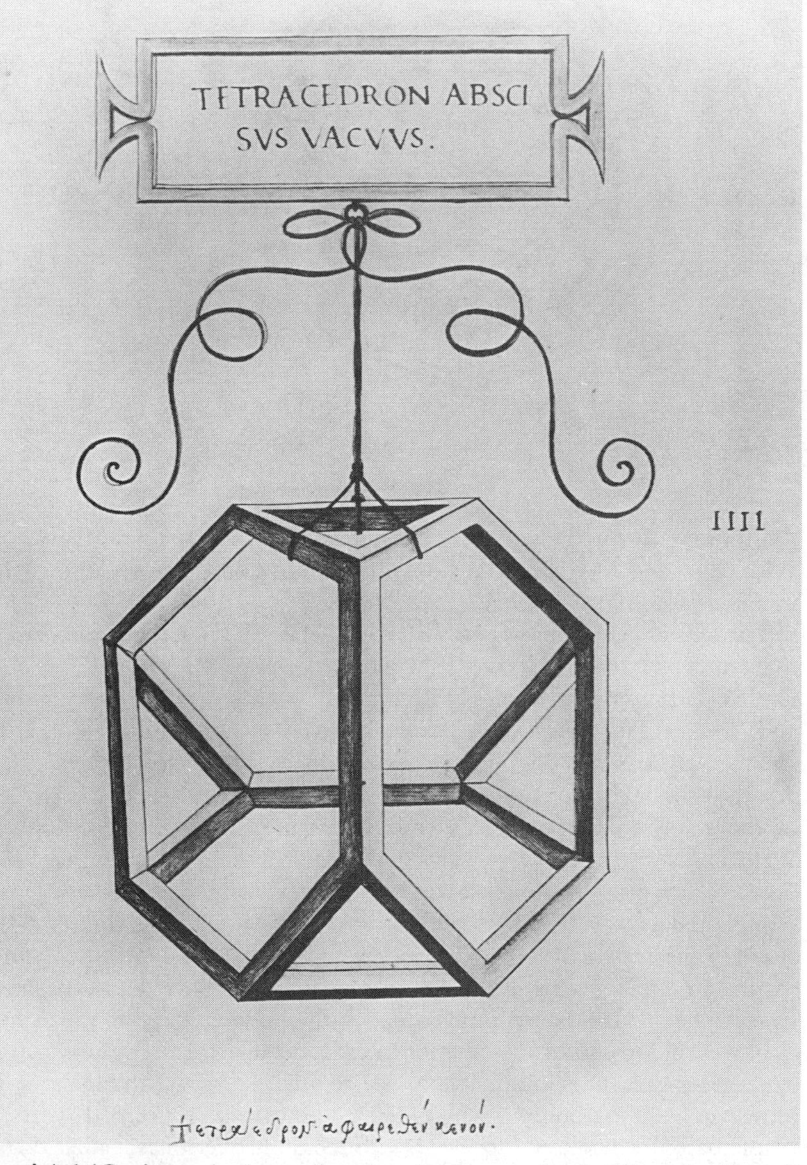

Figure A4.6 'Cut' tetrahedron, drawing by Leonardo da Vinci for Pacioli *De Divina Proportione* (Venice, 1509)

which will produce the cube (figure A4.2), and in order to produce the Platonic dodecahedron he adds roof-like shapes to each face of the cube (figure A4.3). These two procedures are described in the *Epitome* in much the same way as they were described in the *Mysterium Cosmographicum*, but in the later work Kepler not only provides illustrations of the processes but also describes and illustrates a further example of the second procedure: showing that if a suitably-proportioned

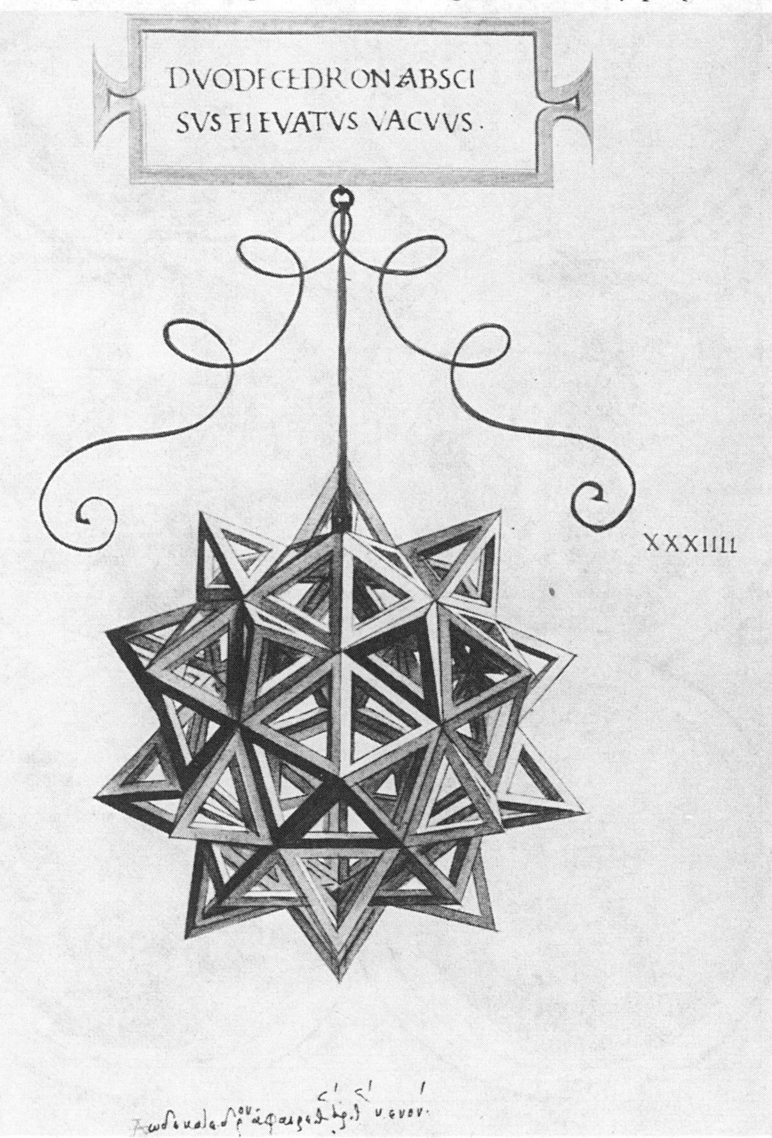

Figure A4.7 'Raised' version of 'cut' dodecahedron, drawing by Leonardo da Vinci for Pacioli *De Divina Proportione* (Venice, 1509)

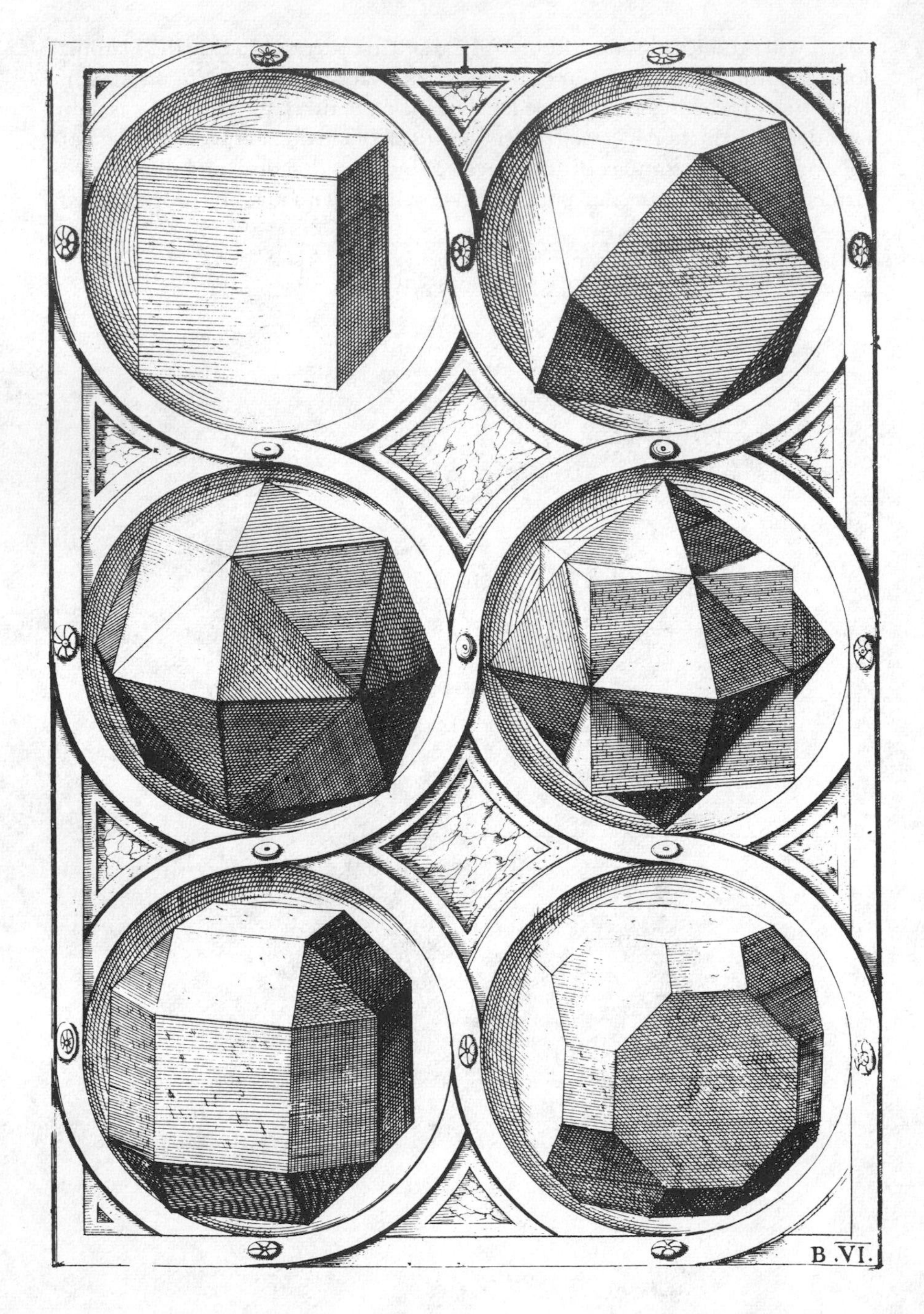

Figure A4.8 Cube and derived solids (Jamnitzer *Perspectiva Corporum Regularium*, Nuremberg, 1568)

Figure A4.9 Dodecahedron and derived solids (Jamnitzer *Perspectiva Corporum Regularium*, Nuremberg, 1568)

square pyramid is added to each face of the cube we obtain the rhombic dodecahedron (figure A4.11). He points out that the faces of the new figure are equilateral but not equiangular, and that its solid angles are of two different types: four faces of the new solid meeting at each of the vertices corresponding to the

Figure A4.10 'Raised' tetrahedron, drawing by Leonardo da Vinci for Pacioli *De Divina Proportione* (Venice, 1509)

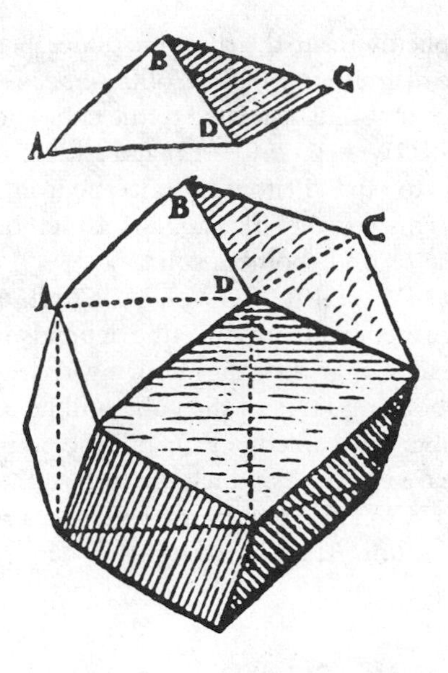

Figure A4.11 Cube and rhombic dodecahedron (*Epitome* IV, p. 461)

apex of one of the added pyramids (such as the point B in figure A4.11), and three faces of the new solid meeting at each of the vertices corresponding to a vertex of the original cube (such as the point D in figure A4.11).

The construction shown in figure A4.11 provides a way of explaining why rhombic dodecahedra, like cubes, can be close-packed. Kepler had discussed this property at length in *De Nive Sexangula* (KGW *4*, pp. 265–6) and mentioned it again, briefly, in *Harmonices Mundi* Book II section V (KGW *6*, p. 68, ll. 25–35). Both discussions mention the fact that when the rhombic solids are packed together, six quadrilinear vertices will fit together to fill space around a point, and four trilinear vertices will do the same. In fact, the packings of the two kinds of polyhedra are mathematically equivalent: the six square pyramids added to the faces of the cube in figure A4.11 will form another cube of the same size as the inner cube (so the packed rhombic solids may be thought of as an array of cubes, whole cubes alternating with cubes made up of the six added pyramids). Kepler seems to have been aware of this identity when he wrote about the rhombic dodecahedron in the *Epitome*, for he says 'And these 6 pieces (*prismata*) add up to little less (*paulo minus*) than the cube on which they are placed' (KGW *7*, p. 270, l. 8). Since in *Harmonices Mundi* Book II Kepler explained, correctly, that six of the quadrilinear solid angles of the rhombic solid filled space round a point (KGW *6*, p. 68, l. 31), one presumes that the rather disconcerting '*paulo minus*' in the *Epitome* is a slip of the pen, or a literary elegance which seems incongruous to a twentieth-century mathematician.

Kepler's analysis of the packing of rhombic dodecahedra in *De Nive Sexangula* is more diffuse than that given in the later work, but it surely must be seen as representing the process of 'discovery', that is the finding of a description of the angles and faces of the solid which explains its known property of filling space.

Although this description does not explicitly relate the rhombic dodecahedron to the cube in the manner shown in the diagram in Book IV of the *Epitome* (see figure A4.11), Kepler does mention that the solid 'is related to the cube and the octahedron' (*De Nive Sexangula*, p. 7, KGW *4*, p. 266, l. 13) and adds that the rhombic triacontahedron is related to the Platonic dodecahedron and icosahedron (KGW *4*, p. 266, l. 14). This last remark suggests a method by which Kepler may have discovered the second rhombic solid.

As Kepler himself remarks (*Epitome* IV, p. 461, KGW 7, p. 270, ll. 6–7), the construction of the rhombic dodecahedron from the cube depends upon adding pyramids whose heights are such that the neighbouring triangular faces of the pyramids added to neighbouring faces of the cube will lie in the same plane. Figure A4.12 shows a cube, with an edge PQ; and the points A and B which are the apices of the square pyramids which are to be added to the faces whose centres are the points X, Y. We require that the faces APQ and BPQ be coplanar, that is, that the line ANB be straight.

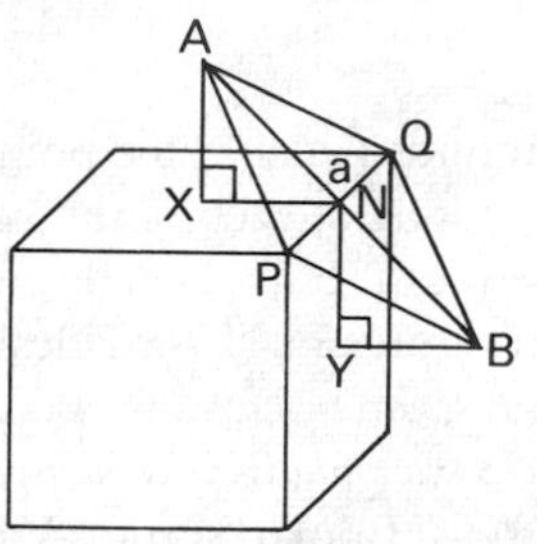

Figure A4.12 Cube to rhombic dodecahedron

If ANB is a straight line, then $\angle$ANB = 180°.

From figure A4.12,

$$\angle ANB = \angle ANX + \angle XNY + \angle YNB$$

Since the solid figure is a cube,

$$\angle XNY = 90°,$$

and, by symmetry, $\angle$ANX = $\angle$YNB.

Therefore, $\angle$ANX = $\angle$YNB = 45°.

Therefore, triangles AXN and BYN are isosceles right-angled triangles.

If the side of the original cube is $2a$, it is clear from figure A4.13, which shows the plane XPQ, that, by symmetry, XN = a.

Since triangle AXN is isosceles, it is clear from figure A4.14, which shows the plane AXN, that AX = XN = a, and by Pythagoras' Theorem, AN = $\sqrt{2}\,a$.

From figure A4.12 we see that the ratio of the diagonals of the rhombic faces of the solid formed by adding pyramids to the cube is the ratio AN : NQ, i.e. $\sqrt{2} : 1$.

The proof has been shortened by appeals to symmetry, but apart from such appeals there is nothing in the proof that cannot be derived from

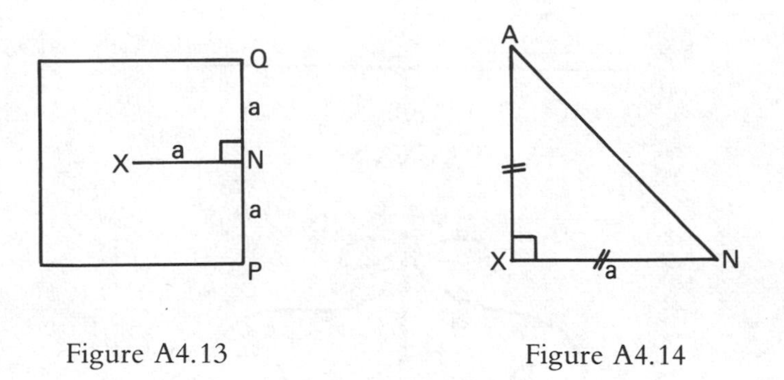

Figure A4.13

Figure A4.14

Euclid, and Kepler's repeated failure to give the ratio which describes the shape of the faces of the solid presumably merely indicates that he expected readers to be able to work it out for themselves. The ratio which describes the shape of the rhombic faces of the triacontahedron can be found in a similar way, by considering pyramids placed on neighbouring faces of a Platonic dodecahedron. The ratio turns out to be $(\sqrt{5} + 1) : 2$.

It seems very likely that Kepler discovered the rhombic triacontahedron by considering exactly this process of adding pyramids to a Platonic dodecahedron. As we have seen, in the *Mysterium Cosmographicum*, and again in *Harmonices Mundi* Book V, he classes the cube, the Platonic dodecahedron and the tetrahedron together as 'primary' figures, one of their 'primary' characteristics being that they have the simplest kind of solid angle, one at which only three planes meet (*Mysterium Cosmographicum*, Ch. III, KGW *1*, p. 29, ll. 22–3). This is a perfectly valid mathematical similarity, and it allows rhombic solids to be constructed from each of the 'primary' polyhedra: Kepler shows in *Harmonices Mundi* Book II section XXVII that no more than three obtuse angles of rhombic faces can meet at a vertex of a rhombic solid, so we clearly could not apply the construction of figure A4.12 to, say, the octahedron, for which four obtuse angles would have to meet at the vertices corresponding to the vertices of the underlying solid. It is not clear whether Kepler realised that it was this property of the vertices which was crucial in constructing rhombic solids to correspond to the 'primary' Platonic figures, but since he lays considerable stress upon the distinction between 'primary' and 'secondary' solids it seems natural to suppose that if he noticed that a rhombic solid could be constructed from one of the primary solids he would try applying a similar construction to the other primary solids. Applying it to the Platonic dodecahedron would have given him the rhombic triacontahedron described in *Harmonices Mundi* Book II and *Epitome* Book IV. Applying it to the third primary solid, the tetrahedron, gives Kepler's remaining rhombic solid: the cube (as can be seen from figure A4.2, if we imagine the pyramids being added to the tetrahedron rather than being taken away from the cube).

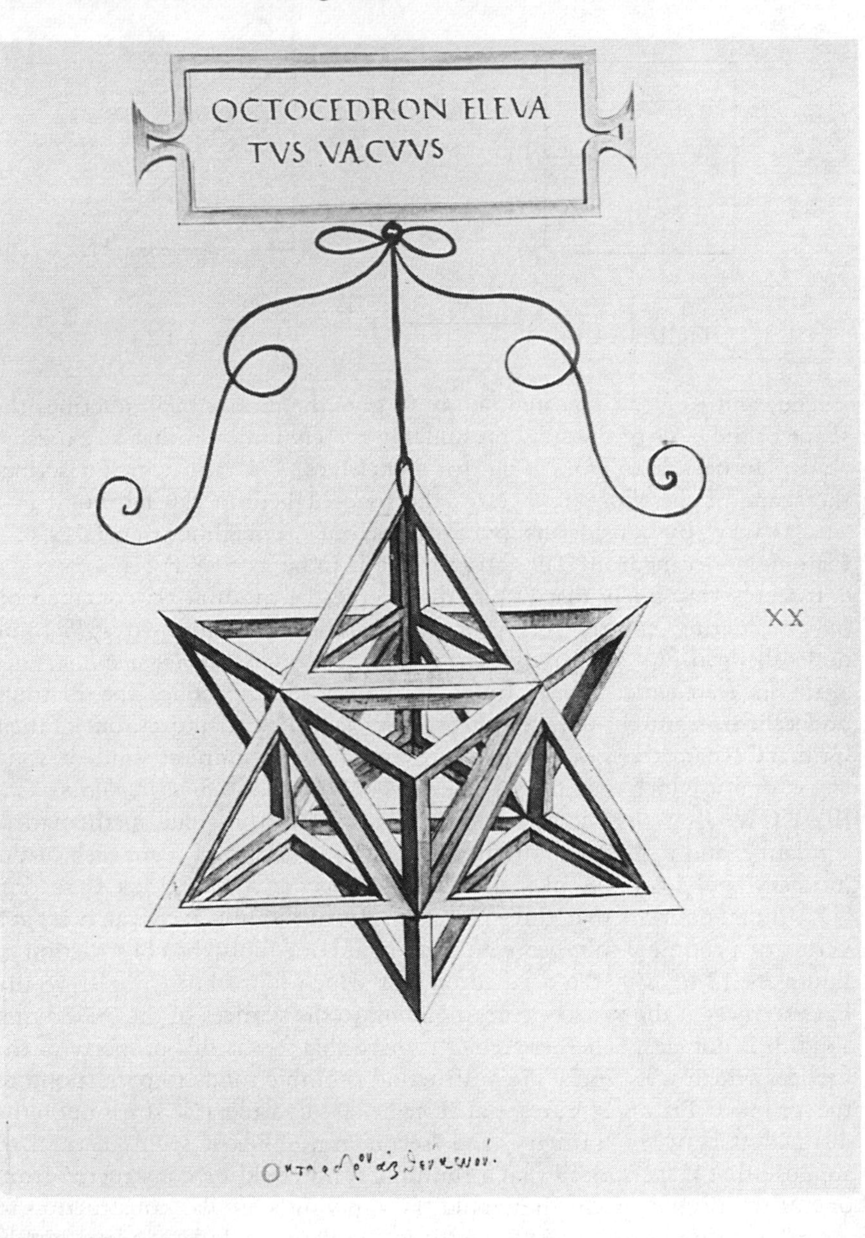

Figure A4.15 'Raised' octahedron, drawing by Leonardo da Vinci for Pacioli *De Divina Proportione* (Venice, 1509)

The orbs associated with the rhombic solids

When we consider associating spheres with the rhombic solids, we come across a possible reason for Kepler's regarding the rhombic dodecahedron as related to the Platonic cube and octahedron, and the rhombic triacontahedron as related to the Platonic dodecahedron and icosahedron.

It is clear from figures A4.11 and A4.12 that the six trilinear vertices of the rhombic dodecahedron lie at the vertices of a cube whose side is the short diagonal of the rhombic faces of the solid. The long diagonals of the faces can be seen to be the edges of an octahedron drawn outside the cube, oriented in the same way as the octahedron drawn inside the cube in figure A4.4. The vertices of this octahedron are the apices of the pyramids that were added to the cube to form the rhombic solid.

The relationship of the rhombic triacontahedron to the Platonic dodecahedron and icosahedron is similar: the short diagonals of the rhombi form the edges of a Platonic dodecahedron, and the long diagonals form the edges of an icosahedron. (Unconvinced readers are advised to make themselves models of the solids.)

The third rhombic solid, the cube, is related to the tetrahedron, or rather to two interlaced tetrahedra. The edges of one tetrahedron are formed by the diagonals shown in figure A4.2, the edges of the other are formed by the remaining diagonals. A line in *De Nive Sexangula* appears to refer to the resultant pair of interlaced tetrahedra '(for the cube is the third, related to two tetrahedra fitted together)' (*De Nive Sexangula*, p. 7, KGW 4, p. 266, ll. 14–15). The discovery of this solid is usually ascribed to Kepler, for example by Coxeter (1975), though there is a picture of it in Pacioli's *De Divina Proportione* (Venice, 1509), where it is constructed as a 'raised octahedron' (see figure A4.15).

Since diagonals of the rhombic faces form edges of regular polyhedra, it is clear that there are three spheres associated with each of the rhombic solids.
(1) The insphere, which touches the solid at the centre of each of its faces. (Since these centres are the midpoints of the edges of the Platonic solid from which the rhombic solid has been constructed, this insphere of the rhombic solid is the same as the midsphere of the inner Platonic solid.) We shall call the radius of this sphere r_i.
(2) The sphere we shall call the 'inner circumsphere', namely the sphere which passes through the vertices of the Platonic solid whose edges are the short diagonals of the rhombic faces. (This sphere is the circumsphere of the Platonic solid from which the solid has been constructed.) We shall call the radius of this sphere r_{ic}.
(3) The sphere we shall call the 'outer circumsphere', namely the sphere which passes through the vertices of the Platonic solid whose edges are the long diagonals of the rhombic faces. We shall call the radius of this sphere r_{oc}.

It is clear that in order to calculate the radii of the three spheres associated with each of the rhombic solids we only need to be able to calculate radii of circumspheres, inspheres and midspheres of the appropriate Platonic solids, and scale the results according to whether the side of the Platonic solid is the short or the long diagonal of the rhombic face. Kepler showed how to calculate the radii of circumspheres and inspheres of the Platonic solids in Chapter XIII of the *Mysterium Cosmographicum*, giving a table of numerical values, to three or four figures, at the end of the chapter. Caspar's note supplies the surds from which these values have been calculated (KGW *1*, p. 46 and p. 425).

Kepler appears to ignore midspheres, though he has in fact calculated the radius of the midsphere of the octahedron when he considers 'the sphere in the square of the octahedron' as an alternative method of finding the aphelion sphere of Mercury (see Chapter III above). However, the relation of the midsphere to the circumsphere is very simple, as can be seen from figure A4.16 in which O is the centre of both spheres and N the centre of an edge. By symmetry, it is clear that $\angle ONA = 90°$.

Let the radius of the circumsphere, $OA = R_c$, and the radius of the midsphere, $ON = R_m$, and let the side of the Platonic solid be $2a$.

Then $AN = a$, and in the plane ONA we have the right-angled triangle ONA shown in figure A4.17.

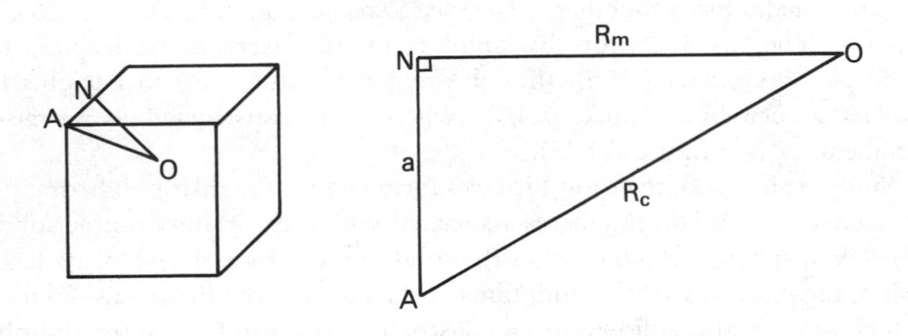

Figure A4.16 Figure A4.17

From triangle ONA, by Pythagoras' Theorem,

$$R_m^2 = R_c^2 - a^2.$$

Figure A4.16 shows a cube only because it is relatively easy to draw: the relation

$$R_m^2 = R_c^2 - a^2$$

holds for all the regular polyhedra. The work in Chapter XIII of the *Mysterium Cosmographicum* will therefore enable us to calculate all the radii we require for the rhombic solids. However, since some of Kepler's numbers are given to only three figures, Caspar's exact surds have been used instead. After a certain amount of manipulation, to make the side of the solid unity in each case, they give the values shown in table A4.1. The gaps in the table indicate values we shall not need to use. Converting the surds into numerical values gives the results shown in table A4.2. The values in these tables can be used to calculate the radii of the three spheres associated with each of the rhombic solids.

Table A4.1 *Radii of spheres associated with the Platonic solids*

solid	R_c radius of circumsphere	R_i radius of insphere	R_m radius of midsphere
cube	$\frac{1}{2}\sqrt{3}$	$\frac{1}{2}$	$\frac{1}{2}\sqrt{2}$
tetrahedron			
dodecahedron	$\frac{1}{8}(3+\sqrt{5})\sqrt{6(3-\sqrt{5})}$		$\frac{1}{2}\sqrt{\frac{1}{2}(7+3\sqrt{5})}$
icosahedron	$\frac{1}{40}(5+\sqrt{5})\sqrt{10(5-\sqrt{5})}$		
octahedron	$\frac{1}{2}\sqrt{2}$		

Table A4.2 *Radii of spheres associated with the Platonic solids*

solid	R_c radius of circumsphere	R_i radius of insphere	R_m radius of midsphere
cube	0.866 025	0.500 00	0.707 107
tetrahedron			
dodecahedron	1.401 259		1.309 017
icosahedron	0.951 057		
octahedron	0.707 107		

(a) The rhombic dodecahedron, taking the length of the short diagonal of the face as unity.

r_i = R_m of cube whose side is 1.
= 0.707 107
r_{ic} = R_c of cube whose side is 1.
= 0.866 025
r_{oc} = R_c of octahedron whose side is $\sqrt{2}$.
= $\frac{1}{2}\sqrt{2} \times \sqrt{2}$
= 1.000 000

(b) The rhombic triacontahedron, taking the length of the short diagonal of the face as unity.

r_i = R_m of Platonic dodecahedron whose side is 1.
= 1.309 017
r_{ic} = R_c of Platonic dodecahedron whose side is 1.
= 1.401 259
r_{oc} = R_c of icosahedron whose side is $\frac{1}{2}(\sqrt{5}+1)$
= 0.951 057 × 1.618 034
= 1.538 843

(c) The cube, taking the side of the face as unity.

$$r_i = R_i \text{ for cube} = 0.500\,000$$

$$r_{ic} = r_{oc} = R_c \text{ for cube} = 0.866\,025.$$

The rhombic solids and the moons of Jupiter

Kepler suggested in his *Dissertatio cum Nuncio Sidereo* (Prague, 1610) that the spacing of the orbits of the newly-discovered moons of Jupiter might be derived from the rhombic solids (KGW 4, p. 309, l. 35 ff, see Chapter IV above). At that time, no measurements of the diameters of the orbits were available, but when Kepler came to write Book IV of the *Epitome* (Linz, 1620), the diameters of the orbits had been found to be in the ratios 3 : 5 : 8 : 13, according to Mayr, and 3 : 5 : 8 : 14, according to Galileo.

The passage in the *Dissertatio* is necessarily vague, but in the *Epitome* Kepler suggests that the rhombic dodecahedron should be placed between the innermost moons, the radii of whose orbs are in the ratio 3 : 5 (KGW 7, p. 318, ll. 41–2). He does not say which orbs of the polyhedron should be considered in the comparison and he does not give any values of radii. He is similarly vague in placing the rhombic triacontahedron between the next pair of moons, the radii of whose orbs are in the ratio 5 : 8, and the cube between the outermost pair (KGW 7, p. 318, ll. 42–4). All the possible ratios are shown in table A4.3.

Table A4.3 *Orbs of rhombic solids and orbs of the moons of Jupiter*

solid	r_{ic}/r_i	r_{oc}/r_i	r_{oc}/r_{ic}	observed ratio
rh. dodecahedron	1.225	1.414	1.154	1.667
rh. triacontahedron	1.070	1.176	1.098	1.600
cube	1.732	1.732	1.000	1.625 Mayr
				1.725 Galileo

The values in this table make it clear that, although the theory may explain the number of the moons, it gives only a very rough account of the spacing of their orbits. They do, however, suggest that Kepler was inclined to accept Galileo's value for the radius of the outermost orbit rather than Mayr's, since he has put the solids in an order which assumes that the outermost space is the largest one. Nevertheless, Kepler does not mention that he has done this, and he does not present any mathematical reasons which might justify his ordering of the solids as he justified his ordering of the Platonic solids in the *Mysterium Cosmographicum* (see Chapter III above). This rather perfunctory treatment of his theory is presumably Kepler's oblique method of passing an unfavourable judgement on it, a judgement which no doubt accounts for the fact that he presents no archetype for the satellite system of Jupiter in *Harmonices Mundi* Book V, despite the fact that the rhombic solids had been described in Book II.

After this unsatisfactory episode concerning the rhombic solids, Kepler goes on to show that the observed diameters of the satellite orbits and the observed periods of the orbital motions agree very well with his third law (*Epitome* IV, pp. 554–5, KGW 7, p. 318, l. 44 to p. 319, l. 3).

Notes

Notes to Introduction

1 These are promised for future volumes of the Complete Works which are being published by the Bavarian Academy of Sciences (*Johannes Kepler Gesammelte Werke*, ed. Caspar *et al.* 1938–).
2 *Mysterium Cosmographicum*, Frankfurt, 1621, New Dedicatory Letter, KGW *8*, p. 9, l. 25 ff. See Chapter IV below.
3 Kepler to Herwart von Hohenburg, 26 March 1598, letter 91, ll. 191–2, KGW *13*, p. 193.
4 Kepler to Magini, 1 June 1601, letter 190, ll. 18–21, KGW *14*, p. 173.

Notes to Chapter I Platonic Science

1 *Harmonices Mundi* Book IV, Ch. I, KGW *6*, p. 221. The quotation from Proclus covers about three sides in the Caspar edition. It is the longest of the many quotations from Proclus that occur throughout the *Harmonice Mundi*, most notably on the title pages of Books I, III and IV (see figure 5.1 below).
Although Kepler does not cite any source for his opinion that *Timaeus* is a commentary on *Genesis*, the opinion is not a new one. It is, for example, found in the writings of the Judaeo-Hellenistic theologian Philo of Alexandria (*fl.* A.D. 40).
2 See Chapter V below and Field (1984c).
3 *Republic* VI, 510d–e
4 Euclid trans. Heath (1956), vol. I, p. 155 and note on p. 224.
5 Euclid trans. Heath (1956), vol. I, p. 254.
6 'Les belles lettres', Paris, 1925 (reprinted 1970).
7 *Republic* VII, 534.

Notes to Chapter II The Size of the Universe

1 See Koyré (1957) p. 58.
2 See, for example, *Mysterium Cosmographicum*, Preface, KGW *1*, p. 9, l. 35; and *Epitome Astronomiae Copernicanae*, Book IV, Part I, first query, KGW 7, p. 258.

3 *De Stella Nova*, Ch. XXI, KGW *1*, p. 253, l. 15.
4 See Koyré (1957) p. 59 ff for a full discussion of Kepler's argument.
5 *De Stella Nova*, Ch XXI, KGW *1*, p. 253, l. 2.
6 *De Revolutionibus*, Bk I, Ch. 8, f6v, NKG II, p. 19, l. 6.
7 KGW *1*, pp. 393–9 and Kepler trans. Field and Postl (1977).
8 KGW *1*, pp. 149–356.
9 *De Stella Nova*, Ch. XXI, KGW *1*, p. 253 ff.
10 *Epitome*, Book I, Linz, 1618, KGW 7, p. 44, ll. 1–6.
11 Prague, 1602, ed. J. Kepler, pp. 481–2.
12 *Epitome*, Book IV, KGW 7, p. 289.
13 *Epitome*, Book I, Part II, KGW 7, p. 42.
14 Koyré notes that Kepler and his opponents shared the belief that if the Universe were infinite the distribution of stars must be uniform, see Koyré (1957) pp. 61–2.
15 KGW 4, pp. 263–80 and Kepler trans. Hardie (1966).
16 Ch. XXI, KGW *1*, p. 256.
17 Book I, Part II, KGW 7, p. 44.
18 *Almagest*, Book I, Ch. 5.
19 *Progymnasmata*, Prague, 1602, pp. 481–2.
20 Misprinted as 34,077,066$\frac{2}{3}$ in *De Stella Nova*, 1606, and in KGW 1, p. 235.
21 *Epitome*, Book IV, Part I, 1620, KGW 7, p. 286.
22 *De Stella Nova*, Ch. XVI, KGW *1*, p. 234, l. 27 ff.

Notes to Chapter III Mysterium Cosmographicum 1596

1 *De Revolutionibus*, Dedicatory letter, NKG, vol. II, p. 6.
2 Reproduced in facsimile, NCG I, f 9v.
3 The work was written as a critique of Cardano's *De Subtilitate* and is in a similar meandering style.
4 *Mysterium Cosmographicum*, Ch. I, KGW *1*, p. 18.
5 This matter is considered more fully in Field (1979b).
6 The second and third editions of the *Narratio Prima* – Basel, 1541 and 1566 – contain no diagrams at all, and Maestlin says in the preface to his own edition that he has added diagrams that were necessary to help the reader to follow Rheticus' proofs (KGW *1*, p. 85, l. 32). I have not been able to inspect a first edition of Rheticus' work.
7 This is described more fully in Field (1984b).
8 *Mysterium Cosmographicum*, Preface, KGW *1*, p. 11, l. 31 ff.
9 *Mysterium Cosmographicum*, Preface, KGW *1*, p. 35, l. 35 ff.
10 *De Stella Nova*, Chapter VI, KGW *1*, p. 179.
11 *Harmonices Mundi* Book II, sect. XXV, KGW *6*, p. 81, l. 21.
12 *Mysterium Cosmographicum*, Preface, KGW *1*, p. 13, l. 6.
13 *Harmonices Mundi* Book II, sect. XXV, KGW *6*, pp. 80–2.

14 See *Mysterium Cosmographicum*, Chapter I, KGW *1*, p. 15, l. 10 ff, Westman (1973) and Jardine (1979).

15 Changing ideas of the size of the planetary system are discussed in Van Helden (1982, 1985).

16 The relation between Kepler's musical and astrological theories is discussed more fully in Field (1984c).

17 For a more detailed discussion see Simon (1975, 1979).

18 Simon (1975), p. 446; see also Field (1984c).

19 See Gingerich (1975). We have used the words 'moving forces', but Kepler here speaks of 'moving spirits' (*animae motrices*) which are weaker when further from the Sun, or a single 'moving spirit' which is in the Sun (KGW *1*, p. 70, ll. 20–1). See note on the corresponding passage in the second edition of the *Mysterium Cosmographicum* in Kepler trans. Aiton and Duncan (1981).

20 Kepler's cosmological model now, quite understandably, seems entirely bizarre to most people, including those best able to undertake the task of actually checking its agreement with 'observations', namely mathematical astronomers. This has had the unfortunate effect of giving currency to the assumption that the theory is, at best, only in rough agreement with observation (and is thus different in spirit from Kepler's astronomical theories). This assumption even seems to have crept into the brief description of Kepler's cosmological theory in *The Titius-Bode Law of Planetary Distances* (Nieto 1972), otherwise an excellent and very interesting book, giving a clear and authoritative account of the curious regularities observed in the Solar system and the various attempts to explain them (up to 1972).

Notes to Chapter IV Mysterium Cosmographicum 1621

1 *Harmonice Mundi*, Linz, 1619, Book II, section XXV, KGW *6*, p. 82, l. 10.

2 *Mysterium Cosmographicum*, Frankfurt, 1621, New Dedicatory Letter, KGW *8*, p. 9, l. 21 ff.

3 *Mysterium Cosmographicum*, Preface, KGW *1*, p. 9, l. 33, KGW *8*, p. 23, l. 34.

4 *Mysterium Cosmographicum*, Frankfurt, 1621, New Dedicatory Letter, KGW *8*, p. 9, l. 12 ff.

5 Kepler to Herwart von Hohenburg, 10 February 1605, letter 325, l. 55, KGW *15*, p. 146.

6 KGW *1*, pp. 3–80, KGW *8*, pp. 1–128.

7 *Mysterium Cosmographicum* Chapter XXI, note (1), KGW *8*, p. 119.

8 KGW *8*, p. 28, l. 35, and p. 93, l. 7.

9 *Narratio de observatis a se quatuor Iovis satellitibus erronibus . . .*, Frankfurt, 1611, KGW *4*, p. 320, l. 6 ff.

10 Having worked on the design of telescopes and having observed the Moon with a number of instruments (all optically superior to Galileo's) my impression is that Galileo must have been very careful and very patient to have seen what he claims to have seen as clearly as his account suggests – that is, sufficiently clearly to prove that there were mountains on the Moon. It is, of course, possible that the account is not a true one, that it is to some extent a description of 'thought observations' designed to persuade others of what Galileo himself already believed. Nevertheless, his account tallies very well with what one can see, with patience and practice, and I find in the *Sidereus Nuncius* a sense of immediacy missing in many of Galileo's descriptions of experiments in other works.

11 *Dissertatio*, *Admonitio ad Lectorem*, KGW *4*, p. 286, l. 4.

12 *Dissertatio*, KGW *4*, p. 287, l. 17 ff.

13 Kepler to Galileo, 13 October 1597, letter 76, KGW *13*, p. 144.

14 *Harmonices Mundi* Book II, section XXVII, KGW *6*, p. 83, trans. in Field (1979a).

15 Book IV, Linz, 1621, Part II, p. 554, KGW 7, pp. 318–9.

16 See above and KGW *8*, p. 9, l. 25 ff.

17 For a fuller account of Kepler's concern with numerology, see Field (1984b).

18 See Drake (1978) and Field (1984a).

19 This theory employs the 'observed' values for the thicknesses of the orbs. Thus Kepler's 'theoretical' dimensions contain an observational element, as a counterpart to the Copernican planetary theory used in producing the 'observed' dimensions.

20 *Epitome Astronomiae Copernicanae*, Book IV, Linz, 1620, Part III, p. 589, KGW 7, p. 337.

21 Kepler to Tanckius, 12 May 1608, KGW *16*, p. 163, letter 493, l. 383. An admirably clear account of the usages of the words *sphaera* etc. in Renaissance texts has recently been given by Jardine (1982).

22 Bialas (1971), pp. 99–140, especially p. 127. See also table 4.6 below.

Notes to Chapter V Harmonices Mundi Libri V

1 Kepler to Herwart, 26 March 1598, letter 91, l. 76 ff, KGW *13*, p. 190

2 *Mysterium Cosmographicum*, Frankfurt, 1621, note on title page, KGW *8*, p. 15, l. 16.

3 Kepler's use of *epidromus* as a counterpart to *prodromus* appears to be a solecism, or perhaps an example of dry Tacitean wit? The context makes the meaning entirely clear, but neither Liddell and Scott nor Lewis and Short nor Ducange record any similar meaning for the word.

4 Kepler to Maestlin, 29 August 1599, letter 132, l. 136 ff, KGW *14*, p. 46.

5 Kepler to Herwart, 14 December 1599, letter 148, l. 11 ff, KGW *14*, p. 100.

6 A fuller account of Kepler's rejection of numerology is given in Field (1984b). See also Walker (1978).

7 Proclus trans. Morrow (1970), p. 19, ll. 10–4 and p. 19, l. 17 – p. 20, l. 2. My three dots correspond to 'Etc' in Kepler's text.

8 Euclid trans. Heath (1956), vol. III, p. 461. Euclid's definition of 'rational', given in Book X Definition 4, means that in this case the diameter is to be commensurable with some given magnitude, either in length or in square.

9 6 August 1599, letter 130, l. 379 ff, KGW *14*, p. 31, see below, p. 116.

10 Kepler defines what is meant by an Aspect at the beginning of *Harmonices Mundi* Book IV (KGW *6*, p. 239 f). He first defines a 'configuration' of two bodies as being the angle between the lines joining the bodies to the centre of the Earth, pointing out that in the case of planets this is the angular distance between them, measured along the circle of the Zodiac. An 'effective configuration' (i.e. an Aspect) is one in which 'the rays from the planets make an angle which is such that it stimulates sublunary Nature and the lower faculties of Animate things' (KGW *6*, p. 240, l. 34). Kepler traces his definitions back to the *Tetrabiblos* of Ptolemy.

11 Kepler to Herwart, 30 May 1599, letter 123, ll. 359–416, KGW *13*, pp. 349–50.

12 Kepler to Herwart, 6 August 1599, letter 130, l. 327 ff, KGW *14*, p. 29.

13 Kepler to Wackher von Wackenfels, early 1618, letter 783, ll. 22–3, KGW *17*, p. 254. See also Walker (1978).

14 Kepler is here rejecting the *numeri numerantes* (counting numbers) of the traditional theories in favour of *numeri numerati*, numbers obtained as the measure of quantity (see Walker (1978) and Field (1984b)).

15 KGW *6*, p. 175, l. 17 ff. This latter passage has been analysed in detail by Walker (1978, p. 51 ff).

16 The Political Digression has been discussed in some detail by Nitschke (1973).

17 For further references see Palisca (1961), Dickreiter (1973) and Walker (1967, 1978).

18 See *De Stella Nova*, Prague, 1606, Chapter V, KGW *1*, p. 172 ff and Simon (1975, 1979).

19 Brengger to Kepler, 7 March 1608, letter 480, ll. 6–11, KGW 16 p. 114; Kepler to Brengger, 5 April 1608, letter 488, ll. 8–15, KGW *16*, pp. 137–8. See also Field (1984c).

20 30 May 1599, letter 123, see above p. 114 ff and figures 5.6, 5.7.

21 Kepler is concerned not with individual planets but with the structure of the Solar system, so the properties he chooses to consider are all, strictly speaking, properties of the planetary orbit rather than properties

of the planet itself, such as its diameter. Presumably Kepler himself regarded this point as obvious, since he makes no explicit reference to it.

22 In fact, each can be inscribed within the other, but Kepler apparently prefers to emphasise the hierarchical distinction between 'primary' and 'secondary' solids first described in the *Mysterium Cosmographicum*, see Chapter III above.

23 That is, the star polyhedron with twelve vertices; see diagrams Ss in figure 5.3.

Notes to Chapter VI Conclusions

1 KGW *6*, p. 290 ll. 3–9. Kepler's reference to the theft of golden vessels (*vasa aurea*) from the Egyptians presents a problem for readers of the Authorised Version (1611) or the *New English Bible* (1970). There are three passages in *Exodus* which refer to the theft of articles made of gold (3.22, 11.2 and 12.35–7), and the articles are described as 'jewelry'. However, Luther's Bible (Augsburg, 1534) translates 'Gefäße' ('vessels') in two passages and 'Geräte' ('implements') in the third one, and the Greek Septuagint (Basel, 1545) gives σκευη, which corresponds closely with Kepler's *vasa*, meaning any kind of utensil or implement. There appear to be no alternative readings of the original Hebrew word (כְּלֵי), and modern scholarship, as embodied in the *Jerusalem Bible* (London, 1966) and the *Hebrew and English Lexicon* (Brown, Driver and Briggs, based on Gesenius, OUP, 1907, reprinted 1968), endorses the translations given by the Septuagint, Luther and Kepler. (I am grateful to my father, Professor E. J. Field, for reading the Hebrew sources on my behalf.)

2 KGW *1*, p. 15, l. 5 ff; see also Jardine (1979).

3 *Astronomia Nova*, Heidelberg, 1609, Chapter XIX, KGW *3*, p. 177, l. 37 ff.

4 KGW *6*, p. 374, ll. 17–18. Yates (1972, p. 223) characterises this comment as expressing 'Kepler's disapproval of Fludd's use of mathematical diagrams as "hieroglyphs"', and Fludd himself seems to have been angered by it, as by almost everything else Kepler said about *Utriusque cosmi . . . historia* (see below). However, Kepler's reply to Fludd's complaint begins

> Then, I say, he is more concerned with practical matters, I with Theories; so pictures are useful (*commodae*) for him, theorems for me. I did not insult you, Robert, I attached no 'invidious mark' (*stigma invidiae*) to your work . . . (*Apologia, Ad Analysin X*, KGW *6*, p. 396, ll. 17–19).

And a little later Kepler asks

> Why then do you flare up? Have I reproached your figures as being silly or a disgrace (*ut insulsas, ut dedecus*)? (ibid. l. 28 ff).

After a few more such questions, Kepler explains what he had intended:

> I compared your pictures to my diagrams; I acknowledged that my book was not as elegantly produced as yours (*non aeque atque tuum ornatum esse*), nor would it be to the taste of every reader: I apologised for this deficiency on professional grounds, since I am a mathematician. (ibid, ll. 32–5)

The remainder of Kepler's reply is mainly concerned with explaining what a student of mathematics may hope to find in the *Harmonice Mundi*. It appears that the distinction Kepler wishes to make is purely technical (and perfectly valid). He does not seem to be expressing disapproval. Nor do his remarks seem worthy of the Jungian analysis they have received from some historians (see Yates (1964, p. 440) for references; and Westman (1984) for further comments). They may, on the other hand, perhaps find their proper place in a history of book-illustration, if one comes to be written?

5 HM III, *Proemium*, KGW *6*, pp. 99–101; Camerarius *Libellus Scholasticus*, Basel, 1551, pp. 205–8. Has Fludd not noticed the refutation that follows Kepler's translation of Camerarius' commentary on the Pythagorean *Carmina Aurea*?

6 In fact, this plane cannot pass exactly through the centre of the Earth since the surface of the 'pyramid' appears to be intended to be tangential to the surface of the Earth. The 'pyramid' would thus have to become an infinite cylinder, rather than a finite half-cone as shown, if its base were to pass through the centre. Close inspection of Fludd's plates reveals that the illustrator has done a certain amount of cheating in order to overcome this problem.

Glossary

This glossary explains words which I am aware of having used in a technical sense not found in the Shorter Oxford English Dictionary.

Element of an orbit. The elements of a planetary orbit are six parameters required to describe it completely. They include the eccentricity of the ellipse, the length of its major axis, the inclination of the plane of the orbit to the plane of the ecliptic and the period of the planet.

Homogeneous. At any given time a homogeneous Universe would present the same appearance to every observer, wherever he was situated.

Isotropic. A Universe is said to be isotropic at a given point if to an observer at that point it looks the same in all directions. By extension, the adjective 'isotropic' is sometimes used to mean 'isotropic at all points' (e.g. by Weinberg, 1977, p. 160).

Musical. In Celestial Mechanics, ratios are said to be musical if they can be expressed in terms of small integers, such as 2 : 3. 'Small' is usually deemed to extend to about 7. Examples of such ratios are found among the periods of some of the satellites of Jupiter (Nieto, 1972, p. 43). See also *resonance.*

Nova. Kepler and his contemporaries used the word *nova* in its primary sense of a new (or apparently new) star. In modern astronomy it denotes an object (probably one star in a binary system) which undergoes an outburst that causes a temporary increase in luminosity of about six magnitudes. For a particular object, this type of outburst is repeated at more or less regular intervals. Compare *supernova.*

Rational. A number is said to be rational if it can be expressed as a ratio between two integers, e.g. 17.4 is rational ($\frac{174}{10}$), $\sqrt{2}$ is not. In geometry a magnitude is said to be rational if it can be measured in terms of some given measure. Kepler finds the word 'rational' misleading and prefers to use 'expressible' (see Chapter V above). Euclid, and Kepler, also use the word now rendered by 'rational' to cover what would now be called 'rational in square'.

Resonance. A system of orbits is said to be in resonance if the periods can be combined to give one or more relations of the form

$$a_1n_1 + a_2n_2 + a_3n_3 + \ldots = \text{constant},$$

where n_1, n_2 n_3, etc. are the periods and a_1, a_2, a_3 etc. are small integers (less than about 7). A striking example of resonance is found in the system of satellites of Jupiter, where the periods of Io (n_1), Europa (n_2) and Ganymede (n_3) satisfy the equation

$$n_1 - 3n_2 + 2n_3 = 0$$

to an accuracy of nine significant figures (Nieto, 1972, p. 44).

Supernova. A supernova is a catastrophic outburst which entirely disrupts a star, leaving behind at most only a small core (which becomes a black hole, a neutron star or a pulsar). There are two types of supernova. In the less energetic type, the object rapidly increases in brightness by roughly eleven magnitudes before fading again, and the total energy of the outburst is about a factor 10^5 greater than that of a nova. A star only becomes a supernova once. The 'new star' which appeared in October 1604 was a supernova of the type just described (see Chapter II above).

Note added in proof Supernova 1987A, the first supernova seen close to us since 1604, seems to differ from both the accepted types. The tally of competing explanations is comparable with that in 1604.

Select Bibliography

I have consulted all the works here cited, some repeatedly, but have not found it profitable to refer to every one in my text.

A fuller bibliography, of works published up to 1975 and referring directly to Kepler, will be found in the *Bibliographia Kepleriana* (2nd edition, Munich, 1968) and its Supplement in the Kepler volume of *Vistas in Astronomy* (Beer and Beer 1975).

Aiton, E.J. (1977), 'Johannes Kepler and the *Mysterium Cosmographicum*', *Sudhoffs Archiv*, 62(2), 173–94.

Aiton, E.J. (1981), see Kepler trans. Aiton and Duncan (1981).

Allen, C.W. (1964), *Astrophysical Quantities*, London, 1964.

Archimedes, *The Works of Archimedes*, trans. T. L. Heath, Cambridge, 1912.

Aristotle, *Meteorologica*, with trans. by H.D.P. Lee, London, 1962.

Aristotle, *On the Heavens*, with trans. by W.K.C. Guthrie, London, 1971.

Aristotle, *Physics*, with trans. by P.H. Wicksteed and F.M. Cornford, London, 1970.

Beer, A. and Beer, P. (eds) (1975), *Kepler, Four Hundred Years* (*Vistas in Astronomy*, 18).

Beer, A., Lyttleton, R. and Richter, N. (1963), *The Nature of Comets*, London.

Bialas, V. (1971), 'Die quantitative Beschreibung der Planetenbewegung von Johannes Kepler in seinem handschriftlichen Nachlaβ', *Kepler Festschrift 1971*, Regensburg, 1971.

Bialas, V., and Papadimitriou, E. (1980), 'Materialen zu den Ephemeriden von Johannes Kepler', *Nova Kepleriana*, 7 (New series), Munich, 1980.

Bible trans. Luther, Augsburg, 1534.

Bible trans. Septuagint, Basel, 1545.

Bible Authorised Version, London, 1611.

Bible Jerusalem Bible, London, 1966.

Bible New English Bible, Oxford and Cambridge, 1970.

Boas, M. and Hall, A.R. (1959), 'Tycho Brahe's System of the World', *Occasional Notes of the Royal Astronomical Society*, 3, 253–63.

Bodin, Jean, *Les Six livres de la république*, Paris, 1576.

Bodin, Jean, *De Republica Libri VI*, Paris, 1586.

Brahe, Tycho, *De Mundi Aetherei Recentioribus Phaenomenis*, Uraniborg, 1588.

Brahe, Tycho, *Astronomiae Instauratae Progymnasmata*, ed. J. Kepler, Prague, 1602 (reprinted Brussels, 1969).

Brahe, Tycho, *Tychonis Brahe Opera Omnia*, ed. J.L.E. Dreyer, 15 vols, Copenhagen, 1913–29.

Buchdahl, G. (1973), 'Methodological aspects of Kepler's theory of refraction', *Internationales Kepler Symposium Weil der Stadt 1971*, Hildesheim, 141–67.

Camerarius, Joachim, *Libellus Scholasticus*, Basel, 1551.

Cardanus, Hieronymus, *De Subtilitate Liber XXI*, Nuremberg, 1550.

Caspar, M. (1938), *Kepler*, ed. and trans. C.D. Hellman, London and New York, 1959. (Original German edition 1938.)

Caspar, M. (1940), 'Nachbericht', KGW *6*.

Copernicus, Nicolaus, *De Revolutionibus Orbium Coelestium, Nikolaus Kopernikus Gesamtausgabe*, 2, Munich, 1949. (Abbreviated to NKG II.)

Copernicus, Nicolaus, *De Revolutionibus* (facsimile of the manuscript), *Nicolaus Copernicus Gesamtausgabe*, ed. H. Nobis, 1, Hildesheim, 1974. (Abbreviated to NCG I).

Copernicus trans. Swerdlow, N. (1973), *Commentariolus*, see Swerdlow (1973).

Cornford, F.M. (1937), *Plato's Cosmology*, Cambridge. (A translation of *Timaeus* with commentary.)

Coxeter, H.S.M. (1948), *Regular Polytopes*, London.

Coxeter, H.S.M. (1969), *Introduction to Geometry*, New York. (First edition 1961.)

Coxeter, H.S.M. (1975), Kepler and Mathematics', *Vistas in Astronomy*, 18, 661–70.

Coxeter, H.S.M., Longuet-Higgins, M.S. and Miller, J.C.P. (1953), 'Uniform Polyhedra', *Philosophical Transactions of the Royal Society of London*, A, 246, 401–50.

Descartes, René, *Les Météores*, Leiden, 1637.

Dickreiter, M. (1973), *Der Musiktheoretiker Johannes Kepler*, Berne and Munich.

Drake, S. (1978), *Galileo at Work*, Chicago.

Dreyer, J.L.E. (1953), *A History of Astronomy from Thales to Kepler*, New York. (Originally published as *A History of the Planetary Systems from Thales to Kepler*. 1906.)

Dürer, Albrecht, *Underweysung der Messung mit Zirkel und Richtscheyt*, Nuremberg, 1525.

Düring, I. (1934) see Ptolemy trans. Düring (1934).

Ehrhardt, A. (1945), 'Vir bonus quadrato lapidi comparatur', *Harvard Theological Review*, 38, 177–93.

Euclid ed. and trans. Foix de Candale (1566), *Euclidis Megarensis Mathematici clarissimi Elementa Geometrica, Libris XV . . .*, Paris.

Euclid trans. Heath, T.L. (1956), *The Thirteen Books of Euclid's Elements*, New York, 3 vols. (A reprint of the second edition, Cambridge, 1926.)

Evans, R.J.W. (1973), *Rudolf II and his World*, Oxford.

Field, J.V. (1979a), 'Kepler's Star Polyhedra', *Vistas in Astronomy*, 23, 109–41.

Field J.V. (1979b), 'Kepler's Rejection of Solid Celestial Spheres', *Vistas in Astronomy*, 23, 207–11.

Field, J.V. (1982), 'Kepler's Cosmological Theories: Their Agreement with Observation', *Quarterly Journal of the Royal Astronomical Society*, 23, 556–68.

Field, J.V. (1984a), 'Cosmology in the Work of Kepler and Galileo', *Novità celesti e crisi del sapere* (Galluzzi, P. ed.), Florence, 207–15.

Field, J.V. (1984b), 'Kepler's Rejection of Numerology', *Occult and Scientific Mentalities in the Renaissance* (Vickers, B.W. ed.), Cambridge, 273–96

Field, J.V. (in press), 'A Lutheran Astrologer: Johannes Kepler', *Archive for History of Exact Sciences.*

Fludd, Robert, *Utriusque Cosmi maioris scilicet et minoris metaphysica, physica atque technica historia in duo volumina secundum cosmi differentiam divisa*, Oppenheim, 1617, 1618.

Galilei, Galileo, *Le Opere di Galileo Galilei* (ed. A. Favaro) Florence, 1890–1909. (Reprinted with additions in the 1930s and 1960s.)

Galilei, Vincenzo, *Dialogo di Vincentio Galilei Nobile Fiorentino della Musica Antica et della Moderna*, Florence, 1581.

Gingerich, O. (1975), 'The Origins of Kepler's Third Law', *Vistas in Astronomy*, 18, 595–601.

Goldreich, P. (1965), 'An explanation of the frequent occurrence of commensurable mean motions in the Solar system', *Monthly Notices of the Royal Astronomical Society*, 130, 159–81.

Goldstein, B.R. (1967) 'The Arabic version of Ptolemy's planetary hypotheses', *Transactions of the American Philosophical Society*, 57, 3–16.

Grafton, A. (1973), 'Michael Maestlin's account of Copernican planetary theory', *Proceedings of the American Philosophical Society*, 117, 523–50.

Grünbaum, B. and Shephard, G.C. (1977), 'Tilings by Regular Polygons', *Mathematics Magazine*, 50, 227–47.

Haase, R. (1971), 'Marginalien zum 3. Keplerschen Gesetz', *Kepler Festschrift 1971*, Regensburg, 159–65.

Hall see Boas and Hall (1959).

Hartner, W. (1968), *Oriens Occidens*, Hildesheim.

Heath, T.L. (1913), *Aristarchus of Samos*, Oxford. (Reprinted 1959, 1966.)

Heath, T.L. (1921), *A History of Greek Mathematics*, 2 vols, Oxford. (Reprinted 1960).

Heath, T.L. (1956) see Euclid trans. Heath (1956).

Hellman, C.D. (1975), 'Kepler and Comets', *Vistas in Astronomy*, 18, 789–96.

Hübner, J. (1975), *Die Theologie Johannes Keplers zwischen Orthodoxie und Naturwissenschaft*, Tübingen.

Jamnitzer, Wentzel, *Perspectiva Corporum Regularium*, Nuremberg, 1568.

Jardine, N. (1979), 'The Forging of Modern Realism: Clavius and Kepler against the Sceptics', *Studies in the History and Philosophy of Science*, 10, 141–73.

Jardine, N. (1982), 'The Significance of the Copernican Orbs', *Journal for the History of Astronomy*, 13, 167–94.

Jardine, N. (1984), *The Birth of History and Philosophy of Science: Kepler's Defence of Tycho against Ursus, with Essays on its Provenance and Significance*, Cambridge, 1984.

Kepler, Johannes, *Johannis Kepleri Opera Omnia* (ed. C. Frisch) Frankfurt, 1865. Abbreviated to KOF.

Kepler, Johannes, *Johannes Kepler Gesammelte Werke*, (ed. Walther von Dyck, Max Casper *et al.*), Munich, 1938–. Abbreviated to KGW.

Kepler trans. Aiton, E.J. and Duncan, A.M. (1981), *Johannes Kepler. Mysterium Cosmographicum. The Secret of the Universe.* The second edition of Kepler's work, reprinted with translation by A.M. Duncan and introduction and commentary by E.J. Aiton, New York, 1981.

Kepler trans. Caspar, M. (1973), *Weltharmonik*, Munich. (First edition 1939.)

Kepler trans. Field, J.V. and Postl, A. (1977), 'Bericht vom Newen Stern', *Vistas in Astronomy*, 20, 333–9.

Kepler trans. Hardie, C. (1966), *The Six-Cornered Snowflake*, Oxford.

Kepler trans. Rosen, E. (1965), *Kepler's Conversation with Galileo's Sidereal Messenger*, New York and London.

KGW see Kepler *Johannes Kepler Gesammelte Werke*.

King-Hele, D. (1975), 'From Kepler's Heavenly Harmony to Modern Earthly Harmonics', *Vistas in Astronomy*, 18, 497–516.

Klein, U. (1971), 'Johannes Keplers Bemühungen um die Harmonieschriften des Ptolemaios und Porphyrios', *Johannes Kepler Werk und Leistung*, Linz, 51–60.

KOF see Kepler *Johannis Kepleri Opera Omnia*.

Koyré, A. (1957), *From the Closed World to the Infinite Universe*, Baltimore.

Koyré, A. (1961), *La Révolution astronomique*, Paris.

Mahnke, D. (1937), *Unendliche Sphäre und Allmittelpunkt*, Halle. (Reprinted Stuttgart 1966.)

Maier, Michael, *Atalanta Fugiens*, Oppenheim, 1617.

Message, P.J. (1982), 'Asymptotic Series for Planetary Motion in Periodic Terms in Three Dimensions', *Celestial Mechanics*, 26, 25–39.

Michel, P.-H. (1962), *La Cosmologie de Giordano Bruno*, Paris.

Mittelstrass, J. (1973), 'Wissenschaftliche Elemente der Keplerschen Astronomie', *Internationales Kepler Symposium Weil der Stadt 1971* (Krafft, F., Meyer, K. and Sticker, B. eds), Hildesheim, 3–27.

Monteverdi, Claudio, *Basso continuo del quinto libro de li Madrigali a cinque*, Venice, 1605.

Morrow (1970) see Proclus trans. Morrow (1970).

Morrow, G.R. (1972), 'Proclus', *Dictionary of Scientific Biography* (Gillispie, C. *et al.* eds), New York.

Neugebauer, O. (1968), 'On the Planetary Theory of Copernicus', *Vistas in Astronomy*, 10, 89–103.

Nieto, M.M. (1972), *The Titius-Bode Law of Planetary Distances*, Oxford.

Nitschke, A. (1973), 'Keplers Staats- und Rechtslehre', *Internationales Kepler Symposium Weil der Stadt 1971* (Krafft, F., Meyer, K. and Sticker, B. eds), Hildesheim, 407–24.

Ovenden, M. (1975), 'Bode's Law – Truth or Consequences?', *Vistas in Astronomy*, 18, 473–96.

Pacioli, Luca, *De Divina Proportione*, Venice, 1509. (Reprinted Milan, 1956, with facsimiles of Leonardo da Vinci's original gouache drawings.)

Palisca, C. (1961), 'Scientific Empiricism in Musical Thought', *Seventeenth Century Science and the Arts* (Rhys, H.H. ed.), Princeton, 91–137.

Pappus of Alexandria, *Collectio* (Commandino, F. ed.), Pesaro, 1588. See also Thomas (1939).

Pedersen, O. (1974), *A Survey of the Almagest*, Odense.

Plato, *The Republic*, trans. B. Jowett, Oxford, 1908.

Plato, *The Republic*, with trans. by P. Shorey, 2 vols, London, 1978, 1980. (First published in 1935.)

Plato, *Timaeus* see Cornford (1937).

Plato, *Platon Oeuvres Complètes*, tom. X, *Timée – Critias*, texte établi et traduit par Albert Rivaud, Paris, 1970. (Copyright date 1925.)

Proclus, *Commentariorum Procli editio prima quae Simonis Grynaei opera addita est Euclidis elementis graece editis*, Basel, 1533.

Proclus, *Procli Diadochi Lycii in Primum Euclidis Elementorum Commentariorum Libri IV a Francisco Barocio Patritio Veneto Editi*, Padua, 1560.

Proclus, *Commentary on the First Book of Euclid's Elements* see Proclus trans. Morrow (1970).

Proclus, *Commentary on Plato's Timaeus* see Proclus trans. Festugière (1966).

Proclus, *Hypotyposis Astronomicarum Positionum*, ed. Grynaeus, Basel, 1540; with trans. by Halma, Paris, 1820; with trans. by Manitius, Leipzig, 1909.

Proclus trans. Festugière (1966), *Commentaire sur le Timée,* traduction et notes par A.J. Festugière, 5 vols, Paris, 1966–69.

Proclus trans. Morrow (1970), *A Commentary on the First Book of Euclid's Elements*, trans. G.R. Morrow, Princeton, 1970.

Ptolemy, Claudius, *Almagest*, with trans. by Halma, Paris, 1813–16.

Ptolemy, Claudius, *Hypotheses*, with trans. by Halma, Paris, 1819. See also Goldstein (1967).

Ptolemy, Claudius, *Tetrabiblos*, with trans. by Robbins, London, 1971.

Ptolemy trans. Düring (1934), *Ptolemaios und Porphyrios über die Musik*, von Ingemar Düring. (German trans. of the *Harmonica* of Ptolemy with notes based on the commentary of Porphyry.) Göteborg, 1934. (Göteborgs Högskolas årrskrifft, bd. 40, no 1.)

Ptolemy trans. Gogava (1562), *Harmonica*, trans. Antonio Gogava, Venice, 1562.

Ptolemy trans. Manitius (1963), *Handbuch der Astronomie*, trans. K. Manitius. (German trans. of the *Almagest.*) Leipzig, 1963.

Reinhold, Erasmus, *Tabulae Prutenicae*, Tübingen, 1551.

Rheticus, Georg Joachim, *Narratio Prima*, Danzig, 1540. Second edition Basel, 1541. Third edition Basel, 1566 (with second edition of Copernicus *De Revolutionibus*). Fourth edition Tübingen, 1596 (with Kepler *Mysterium Cosmographicum*).

Righini Bonelli, M.L. and Shea, W.R. (1975) (eds), *Reason, Experiment and Mysticism in the Scientific Revolution*, New York, 1975.

Rosen, E. (1965). See Kepler trans. Rosen (1965).

Rosen, E. (1975), 'Kepler and the Lutheran attitude towards Copernicanism', *Vistas in Astronomy*, 18, 317–37.

Rosen, E. (1984), 'Kepler's attitude toward astrology and mysticism', *Occult and Scientific Mentalities in the Renaissance* (Vickers, B.W. ed), Cambridge.

Sambursky, S. (1962), *The Physical World of Late Antiquity*, London.

Scaliger, Julius Caesar, *Exercitationes Exotericae*, Paris, 1557.

Shephard (1977). See Grünbaum and Shephard (1977).

Simon, G. (1975), 'Kepler's Astrology: the Direction of a Reform', *Vistas in Astronomy*, 18, 439–48.

Simon, G. (1979), *Kepler astronome astrologue*, Paris.

Swerdlow, N. (1973), 'The Derivation and First Draft of Copernicus' Planetary Theory: A Translation of the *Commentariolus* with Commentary', *Proceedings of the American Philosophical Society*, 117, 423–512.

Thomas, I. (1939), *Greek Mathematical Works. Selections illustrating the history of Greek Mathematics*, 2 vols, London, 1939–41.

Van Helden, A. (1982), 'Galileo on the sizes and distances of the planets', *Annali dell'Istituto e Museo di Storia della Scienza di Firenze*, Anno 7 (2), 65–86.

Vlastos, G. (1975), *Plato's Universe*, Seattle.

Walker, D.P. (1967), 'Kepler's Celestial Music', *Journal of the Warburg and Courtauld Institutes*, 30, 228–50.

Walker, D.P. (1978), *Studies in Musical Science in the Late Renaissance*, London.

Weinberg, S. (1977), *The First Three Minutes*, London.

Westman, R.S. (1973), 'Kepler's Theory of Hypothesis and the "Realist" dilemma', *Internationales Kepler Symposium Weil der Stadt 1971* (Krafft, F. Meyer, K. and Sticker, B. eds), Hildesheim, 29–54.

Westman, R.S. (1977), 'Magical Reform and Astronomical Reform: The Yates Thesis Reconsidered', *Hermeticism and the Scientific Revolution*, McGuire, J.E. and Westman, R.S., Los Angeles.

Westman, R.S. (1984), 'Nature, art and psyche: Jung, Pauli and the Kepler – Fludd polemic', *Occult and Scientific Mentalities in the Renaissance* (Vickers, B.W. ed.), Cambridge.

Yates, F.A. (1964), *Giordano Bruno and the Hermetic Tradition*, London.
Yates, F.A. (1972), *The Rosicrucian Enlightenment*, London.
Yates, F.A. (1979), *The Occult Philosophy in the Elizabethan Age*, London.
Zarlino, Gioseffo, *Istitutioni Harmoniche*, Venice, 1558.

Addenda to Bibliography

The following works have been published since the completion of the manuscript of my book (June 1984).

Field, J.V. (1984c), 'A Lutheran astrologer: Johannes Kepler', *Archive for History of Exact Sciences*, 31(3), 189–272 [includes an English translation of Kepler's *De fundamentis astrologiae certioribus* (Prague, 1602)].
Field, J.V. (1986), 'Two mathematical inventions in Kepler's *Ad Vitellionem paralipomena*', *Studies in History and Philosophy of Science*, 17(4), 249–268.
Kepler trans. Segonds (1984), *Jean Kepler. Le Secret du Monde* [French translation of *Mysterium Cosmographicum*], trans. and annotations by A. Segonds, Paris: Les Belles Lettres.
Lindberg, D.C. (1986), 'The Genesis of Kepler's Theory of Light: Light Metaphysics from Plotinus to Kepler', *Osiris* (2nd series), 2, 5–42.
Van Helden, A. (1985), *Measuring the Universe*, Chicago and London: Chicago University Press.

Index